I0817607

Gathered

Gathered

ON FORAGING, FEASTING, AND THE SEASONAL LIFE

GABRIELLE CERBERVILLE

HARPER

An Imprint of HarperCollins*Publishers*

This book contains advice and information relating to foraging and consuming foraged plants and fungi. All included information has been personally tested by the author and fact-checked by multiple experts for safety and accuracy. However, please note that this book is not comprehensive—it contains only a small selection of plants and mushrooms worldwide. Always cross-examine identifications and preparation recommendations with multiple trusted sources before consuming any foraged plant or mushroom. Readers should always proceed with caution before ingesting anything that they forage, as some plants and fungi are toxic, and the publisher and author disclaim liability for any damages that may occur as a result of incorrect identification or improper preparation. Make sure that you are clear on whether or not foraging is permitted wherever you intend to harvest.

 For information, address HarperCollins Publishers, 195 Broadway, New York, NY 10007. In Europe, HarperCollins Publishers, Macken House, 39/40 Mayor Street Upper, Dublin 1, D01 C9W8, Ireland.

HarperCollins books may be purchased for educational, business, or sales promotional use. For information, please email the Special Markets Department at SPsales@harpercollins.com.

hc.com

FIRST EDITION

Designed by Bonni Leon-Berman

Illustrations by Madison Memering

Library of Congress Cataloging-in-Publication Data has been applied for.

ISBN 978-0-06-335791-4

25 26 27 28 29 LBC 5 4 3 2 1

For all my teachers,

human, and especially otherwise

Contents

Foreword

THE FIRST TIME I MET GABRIELLE she was playing a mushroom.

She had attached two electrodes to a large hen of the woods, and with the help of some very special witchery (called a PlantWave), we could actually hear the gentle song of the maitake. I was teaching at the Midwest Wild Harvest Festival where Gabrielle was the keynote speaker, showing us all the ways in which her work combines computing skills, music composition, and a deep love of nature. We were hooked.

I knew immediately Gabrielle was a kindred spirit. What I didn't know was that she was so capable of baring her soul and sharing her philosophies in language so lovely and evocative. Like I did, you may find yourself reading this book in bed in the wee hours, promising yourself, "Just one more chapter."

As a forager and an author, I've read a lot of books about wild foods. (Really. A lot.) But *Gathered* is different from the other books on wild food. This book you hold in your hands is unique—part Gabrielle's story, part recipe collection, all the ongoing journey of one woman's intimate relationship with the natural world that surrounds and sustains her. These things are essential to Gabrielle: the natural world provides sustenance and comfort, as do the recipes she shares. And with this sustenance, Gabrielle emerges strong and resilient. She shares her favorite authors with us, and explains how foragers view the world, describing the relationships we cultivate with the landscape, the plants and animals, and with other foragers. She also supplies the reader with plenty of practical information about tools, food preservation, and sustainability.

But this is not a field guide or a cookbook. There's a story here. The book begins with loss. Understandably, Gabrielle is unsure of her path. She learns to rely on the landscapes and plants she loves and the people who love her, and along the way a philosophy emerges, one that encourages all of us to respect and revel in the natural world, to take with gratitude and to remember to give, and to celebrate every season, diving deep into the mud and sticks and prickers because there's something worthwhile underneath.

Gabrielle says she isn't a scientist, but she has an analytical mind and clarity of thought. Yet her writing is never sterile or intimidating. It is, in fact, poetic. Foraging is often a solitary activity, although you may run into some large, furry creatures in your favorite blueberry patch (yes, that happens). But Gabrielle reminds us that we are not alone. You're in the woods with Gabrielle leading the way. Irony alert: turns out Gabrielle grew up in the same small Pennsylvania town where I had a house for thirty years and yet we never met until we'd both moved away. Reading her descriptions of the woods and mountains around that small town brought me back immediately; it made me miss the wintergreen and the blackcaps, and the chanterelles. Sigh.

Gabrielle describes this book as "a love letter of sorts to the observable, inhabitable, edible wild world that has so deeply shaped me." And that's exactly right. It will shape the reader, too, if you take it to heart. You definitely should.

Ellen Zachos,
author of *Backyard Foraging* and *Mythic Plants*

Glossary

Don't Panic!

This glossary includes several botanical (plant) and mycological (fungal) terms and concepts you may be unfamiliar with, but don't stress about them! You'll find this more useful later in your reading for reference, especially when you are interpreting the information included in the glossary. As you continue learning and growing, terms like these will become more useful to you, and will improve the speed and accuracy of your identifications. Keep at it and remember that you can always look stuff up!

PLANT TERMS

Achene: A tiny, dry fruit that usually contains just one seed and does not split open when ripe (for example, the small "seeds" on a strawberry's surface).

Annual: A plant that completes its life cycle in one year (typically sprouting in spring, blooming in summer, and dying in fall).

Basal Rosette: A circular cluster of leaves that grow very close to the ground, all radiating outward from the same point at a plant's base.

Biennial: A plant that completes its life cycle in two years, usually growing leaves the first year and then flowering and producing seeds the second year.

Bract: A small, leaflike structure that appears at the base of a flower (for example, wild carrot).

Catkin: A slim, tube-like cluster of small flowers that typically hangs from branches (commonly seen in trees like willows or birches).

Compound: A leaf that is made up of several distinct parts (leaflets) on a single stem.

Deciduous: Describes trees and shrubs that lose their leaves each fall and grow new ones in the spring.

Dioecious: A plant species in which male and female flowers grow on separate individual plants (one plant makes pollen, another makes seeds/fruit).

Drupe: A fleshy fruit with a single hard seed or "pit" inside (peaches and cherries are well-known drupes).

Fascicle: A bundle of leaves or flowers growing closely together at the base from a twig. The number of needles per fascicle from a pine tree is a key identifying feature.

Hirsute: Covered in thick, coarse hairs.

Inflorescence: A cluster of flowers on a single stalk, often appearing visually as a single flower (for example, dandelion flowers).

Lanceolate: Shaped like a lance tip: long, narrow, and wider in the middle before tapering to a point.

Lenticels: Small, pore-like bumps on a stem or bark that let gases (like oxygen) move in and out of the plant's tissues.

Monoecious: A plant species in which male and female flowers grow on the same individual plant (so one plant can produce both pollen and seeds/fruit).

Ovate: Egg-shaped, usually referring to leaves that are broader at the base and taper toward the tip.

Pappus: A ring of tiny hairs or scales at the top of a seed (in plants like dandelions) that helps it float in the wind.

Perennial: A plant that lives for more than two years, often returning (or staying green) each growing season.

Rhizome: A horizontal, underground stem that can sprout new shoots and roots, helping the plant spread.

Siliques: The slender, elongated seed pods of certain plants in the mustard family (they split open from both sides to release seeds).

Umbel: A cluster of flowers where each flower's stalk starts from the same point, creating an umbrella-like shape (as in Queen Anne's lace).

FUNGI TERMS

Bolete: A type of mushroom that has a thick, spongy layer of pores under its cap instead of thin gills.

Decurrent: Describes gills or ridges on a mushroom that run down part of the stem rather than stopping right at the cap's edge.

Gills: The thin, blade-like structures on the underside of many mushroom caps where spores are produced (they look like the pages of a book, or like a fan).

Hymenium: The underside of a mushroom's cap, typically the spore depositing surface.

Mycorrhizal: Describes a relationship where a fungus lives on or in plant roots, exchanging nutrients in a way that helps both the fungus and the plant.

Parasitic: Living on or in another organism (a host) in a way that harms the host; for mushrooms, this often means infecting and damaging trees (such as honey mushrooms).

Polypore: Fungi that produce "brackets," which grow in shelves, usually from dead or dying wood, and which have many small holes on the underside (pores) rather than gills or teeth.

Reticulation: A net-like pattern or texture found on some mushroom stems (for example, you might see a faint "net" design on porcini, or a more dramatic one in a Frost's bolete).

Rhizomorph: Thick, root-like strands of fungal tissue that transport nutrients and can spread from one spot to another (often found in honey mushrooms).

Saprotrophic: Describes fungi that feed on dead or decaying organic matter (like rotting wood or leaf litter), helping break it down.

Spore Print: The powdery deposit left behind when a mushroom cap is placed underside-down on a surface; it shows the color of the spores.

Stipe: The stem or stalk of a mushroom, which holds up the cap.

Teeth: Spine-like, tubular structures that typically hang downward from the underside of certain fungi, some of which are referred to as "hydnoid" fungi (for example, lion's mane or hedgehog mushrooms).

PARTS OF A FUNGUS

Cap: The top portion of the mushroom, often umbrella- or funnel-shaped. This is usually the most visible part.

Gills: Thin, blade-like structures on the underside of the cap where spores are produced. Not all mushrooms have gills—some have pores or other surfaces instead.

Pores: Tiny holes on the underside of certain mushrooms (like boletes) instead of gills. The pores release spores when the mushroom is mature.

Stipe (Stem): The stalk that supports the cap. Some mushrooms have long stipes, others have very short or nearly absent stipes.

Ring (Annulus): A ring of tissue that sometimes remains on the stipe after the mushroom's partial veil (a protective covering) tears away as the cap opens. Not every mushroom has a ring.

Volva: A cup- or sac-like structure at the base of some mushrooms (especially in the *Amanita* family). It's the remnant of the universal veil that once enclosed the entire young mushroom.

Partial Veil: A temporary "cover" that protects the gills or pores when the mushroom is still developing. Parts of the partial veil can form the ring on the stem once it breaks.

Universal Veil: A thin membrane that surrounds some mushrooms when they are young. As the mushroom grows, it breaks open, sometimes leaving behind patches or warts on the cap and a volva around the base.

Mycelium: The "main body" of a fungus, typically hidden under the soil or within dead wood. It's made of thin, threadlike strands that gather nutrients and water.

PARTS OF A PLANT

Bark: The tough outer covering of the stems and trunks of woody plants (like trees and shrubs). It protects the inside of the plant.

Compound Leaf: A leaf that's divided into smaller leaflets, all attached to a single leaf stem. "Pinnately compound" means the leaflets are arranged on opposite sides of a central stalk, like a feather.

Flowers: The reproductive structures of many plants. Flowers often have colorful petals to attract pollinators. Inside, you'll find male parts (stamens) that produce pollen, and female parts (pistils) that produce seeds.

Fruits: Structures that develop from flowers and contain seeds (for example, a peach, tomato, or acorn). Their purpose is to help a plant spread its seeds.

Leaves: The food-making structures of most plants. They capture sunlight for photosynthesis. Each leaf often has a few key parts:

- **Blade:** The broad, flat part.
- **Margin:** The edge of the leaf (which can be smooth, wavy, or serrated).
- **Midrib:** The main vein running up the center of the leaf.
- **Petiole:** The stalk attaching the blade to the stem.

Nodes: Places on a stem where leaves or branches attach.

Rhizome: A horizontal underground stem that can send out new shoots and roots, allowing the plant to spread.

Roots: The part below ground that anchors a plant and absorbs water and nutrients from the soil.

Seeds: The plant's reproductive units that can grow into new plants if conditions are right (soil, water, and sunlight).

Stem (or Trunk in Trees): The main support structure that holds up leaves and flowers. In trees, it's called a trunk and is covered with bark.

Taproot: A large, central root (like a carrot) that goes straight down, with smaller side roots branching off.

Umbel: A group of short flower stalks that spread out from a single point, making an umbrella-like shape (seen in plants like Queen Anne's lace).

COMMON LEAF SHAPES

Cordate (Heart-Shaped): Looks like a heart, with a notch at the base and a rounded point at the tip.

Elliptical: Shaped like a flattened oval; widest at the middle and tapering at both ends.

Lanceolate: Shaped like a lance or spear: long and narrow, widest near the base, and tapering to a point.

Linear: Very narrow, generally the same width from base to tip, like a blade of grass.

Oblong: Longer than it is wide, with sides that are generally parallel, and ends that are rounded (not pointed).

Orbicular (Circular): Nearly circular in shape, sometimes with a small notch for the petiole.

Ovate (Egg-Shaped): Broadest near the base (where it attaches to the stem) and tapering toward the tip—like an egg on its side.

Peltate (Shield-Shaped): Circular or rounded leaf where the petiole (leaf stalk) attaches underneath the center of the leaf, not at the edge.

Reniform (Kidney-Shaped): Similar to heart-shaped but with a shallower notch and a rounder overall form, like a kidney bean.

Sagittate or Hastate (Arrow-Shaped): A roughly triangular leaf with pointed lobes that spread outward at the base, resembling an arrowhead.

Spatulate (Spoon-Shaped): Broad, rounded tip that narrows significantly toward the base, like a spatula or spoon.

COMMON GILL ATTACHMENT TYPES

Adnate: The gills are attached directly to the stem, meeting it at a roughly right angle. They do not run down the stem or have a notch.

Adnexed: The gills are only slightly attached to the stem at the very top, often tapering or narrowing near the point of attachment. This can look like a very small, slanted connection.

Decurrent: The gills run clearly down the stem, so each gill continues along the stem for a short distance.

Emarginate: Similar to sinuate, but the notch is sometimes less pronounced. The gills still make a shallow inward curve close to the stem, rather than attaching at a sharp angle. This term is often used interchangeably with *sinuate*, though sinuate tends to be a deeper notch.

Free: The gills do not touch or attach to the stem at all. There is typically a small gap or "collar" around the top of the stem.

Sinuate (Notched): The gills curve upward near the stem, leaving a small notch or inward curve before attaching. When viewed from the side, there's a distinctive "dip" near the stem.

OTHER GILL FEATURES

Color changes: Gills can change color with age or when bruised, which can be an important ID clue.

Forking: In a few types of mushrooms, the gills may appear to branch or fork near the edge.

Spacing: Gills might be described as "crowded" (very close together) or "distant" (with noticeable space between).

Thickness: Some mushrooms have thick, fleshy gills; others have very thin, papery gills.

Preface

WELCOME TO THE WONDERFUL, wild, edible world. This book is my attempt at teaching you many things; chiefly, the practicalities of foraging, processing, and eating wild food, but additionally, it exists to impart some of the nuance that is somewhat more difficult to include in a traditional field guide or video. As such, I have chosen to write it in the style of a memoir; a love letter of sorts to the observable, inhabitable, edible wild world that has so deeply shaped me.

Wild foodies and naturalists are not trendsetters living on some imaginary cutting edge of sustainability, and we are not fringe extremists. In fact, we are, more often than not, simply people who have a wide variety of experiences and interests. Some of us come by the practice out of necessity or as a way of interacting more with our own culture, drawing on the wisdom of grandmothers, elders, and others who had the good sense to survive together in lean times and thrive together in abundant times. Others are attracted to the idea of self-reliance, taking back agency and power from the commercialized food systems and divesting from capitalism. Still others find their way through a single spark of curiosity, a wonderment that ignites a fire of learning and exploration. It is simply good sense that we should find joy in gathering beautiful treasures from the outdoors and transform them, alchemy-like, into delicious meals, and that we should empower ourselves with the knowledge and wisdom to recognize and respond to the bounty that is already within our reach.

I did not go to school to learn about plants or fungi, and I am not part of a familial or cultural lineage of foragers in living memory. My father is a New Yorker born to Italian and Puerto Rican immigrants, my mother is a proud army brat of German stock, simultaneously from everywhere and nowhere at all. My parents homeschooled my two younger brothers and me, raised us feral, barefoot, and free in the mountains of Milford, Pennsylvania. In our home among the oaks and hemlocks, they sheltered us from city sidewalks, computer games, and MTV. The forest was our home, our teacher, our entertainment. We did not need to be pushed outside to go play; we had to be dragged back *in*. Spring brought newts and frogs to be caught and examined, summers were for swimming in the pond and picking wildflowers, having mud fights and persistent blueberry-purple tongues. Autumn brought leaf piles and campfires with half-burnt hot dogs still cold in the middle, and winter meant sledding down the big hill by the house, ice skating, snow forts, and hideaways.

Forests have since taught me how to live. They have taught me resilience, creative problem-solving, how to be alone with myself, and even how to be together with others. I have stumbled and fought my way through clumsy mistakes, errors in judgment, and learned to contend with the worst parts of myself—my self-centeredness, impatience, and ungraciousness.

I am what is often incorrectly referred to as a "self-taught" forager, inasmuch as I did not apprentice with any specific person. The forest is and has always been my best teacher, but my practice has been honed and shaped by many teachers over the years. I am not self-taught. We learn by paying attention to and practicing what we see and hear, whether it be the words of others or the environments we examine and ingratiate ourselves to. We learn by asking questions, by using our creativity coupled with our memories, filing our experiences away and recombining them to solve problems and make new discoveries about how to live in this world. My body of knowledge is not my own; rather, it is a mosaic of all the people and plants and mushrooms and animals I have learned from over the past many years, and I hope that this book will be a part of your own educational mosaic as you begin to learn not only *what* to see, but *how* to see on your foraging journey. I hope that you will offer what you learn through the years to others, just as I'm doing here.

My own philosophy, which I hope to impart to you, has grown out of others' ideas, central to the way I have come to see the natural world. Chief among those foundational ways of thinking is a concept that, to my mind, all foragers who concern themselves with developing a harmonious, reciprocal relationship with their environment should adopt, known as the Honorable Harvest. The name is attributed to Dr. Robin Wall Kimmerer, distinguished plant ecologist and author of *Braiding Sweetgrass,* but as she reminds us, it did not start with her. The Honorable Harvest is a way of living with nature that is well known to many Indigenous people of this land; a way of centering the pursuit of relationships—with other people as well as with the natural world—as an alternative to the pursuit of individual wealth and exploitation. The Honorable Harvest leads us away from consumerism, instead offering a redirection of purpose and future-mindedness by reminding humanity of our responsibility to return life to the places from which we take life for ourselves. It is not just applicable to foraging, though it has direct ethical considerations; rather, it is an alignment of reciprocity, restraint, gratitude in all things, and respect as a primary way of approaching all our interactions. Among its chief teachings are:

1. **Ask permission.** Plants and mushrooms and animals belong to themselves, and we all belong to a natural ecosystem together, but just as I would ask before digging through your purse for a stick of gum, it is appropriate to ask a tree before you pick its fruit. Even if it feels a bit weird, it's important to acknowledge the "alive-ness" of your nonhuman neighbor, thus creating space to contemplate your choice to harvest or not.

2. **Leave something behind.** When harvesting a native plant or mushroom, forgo pocketing the first or last. This is a good way to exercise restraint, and to ensure that there is always something left behind. Every organism on earth benefits from having some of itself left behind to reseed, respawn, set spores, or provide nourishment for others. Just because something is there, does not mean it is there for you. It is important to note that harvesting invasive species falls under a rather different set of suggestions than harvesting native species. As we strive to maintain ecological balance in our stewardship

role, it is critical to remove invasive species to prevent their spread—this, too, is environmental care. Fortunately for us, many of these are edible, and many of the species that I will introduce you to in the following pages happen to be edible invasives. Take all you can use, and more if you're able. When removing edible invasives, I still like to thank them for feeding me, to appreciate how they have tried to make a home on foreign soil, and remind myself that they didn't bring themselves here.

The work of healing must be led by those who have done the hurting, and in the absence of those people, that work falls to us, their children. We have a responsibility to improve the future by mending the past.

3. Practice gratitude. Practicing gratitude goes beyond saying "thank you" and moving on with your day. It means that we are choosing to enter into a practice of gift-giving, a relationship built on the understanding that we will complete the "circle of gifts" by both receiving and giving. Unless we begin to view nonhumans such as plants, animals, and fungi as our neighbors and relatives, we will never understand how to treat them with respect, learn to meet their needs, or flourish alongside them. Entering into reciprocity with your local flora, fauna, and fungi builds kinship, kindness, and purpose into your life, and makes you a better steward of the land that holds you. When you don't develop a relationship with the organisms that you forage from, a plant is just a plant, and a mushroom is just a mushroom. But more important, you are missing out on the best gift that foraging gives us—an invitation to belong to something much greater and more meaningful than ourselves. As Diné activist Lyla June Johnston reminds us, humans were "born to be givers. We were born to be stewards. We were born to reach out beyond our own species and care for the Earth."

I COULD HAVE WRITTEN ABOUT hundreds of different edible plants and mushrooms, and many of those books already exist (I made a list of some of them, and you can check it out in the back of this book). But I am not primarily a field researcher; I am an artist. I am a storyteller, and the forest I invite you to visit has many stories to tell. This book is not a field guide so much as it is "a guide to fielding," by which I mean that this book is meant to be your companion and mentor as you embark on a journey of experiencing, relating to, and building community with the natural world through the marvelous act of reciprocal participation we call foraging. Through the introductions I have provided to these few yet remarkable edible allies in the botanical and fungal worlds, I hope that you will begin to see with open eyes that we can learn from nature, not just about it. In the Gathering Exercises provided at the end of each chapter, I have endeavored to give you a pathway to honing your senses, to rooting yourself in the natural world not as a mere observer, but as a participant.

Gathered is written to be accessible to beginners (there is so much to learn that I often still feel like a beginner myself). However, in order to make sure you are absolutely confident in your identifications, sometimes we need to use botanical or mycological terms that beginners may be unfamiliar with. If you are just starting your foraging journey, you will be provided with ample opportunity to beef up your vocabulary and hone your senses, but I will try to make sure that I define my terms along the way. As you pick up and absorb essential vocabulary, you will also begin to intuit the critical questions you should be asking and the sensory features of plants and mushrooms you should be paying attention to. In time, it's my hope that by learning how to notice and recognize these features and pair them with relevant descriptive words, you'll be able to navigate field guides geared more strictly toward identification with better fluency, taking your learning far beyond the bounds of this book.

As a first tip I'll say: Be wise, be cautious, and be humble when identifying. Your first idea of an identification may not always be correct. *You will get things wrong.* You will make mistakes. Never cling so tightly to your pride that you cannot unclench your fist or open your mind to receive new information. I don't say these things to frighten you; rather, I hope to instill in you a healthy *respect* for the power of the natural world. Some plants and fungi have delicious gifts to share with you, all have lessons to teach us, but many are not meant for us to consume. You, fellow human, like me, are a part of nature, and as such, you *belong* in respectful communion with your nonhuman relatives, plants, fungi, and animals alike.

Each edible plant or mushroom that I introduce to you will have an anatomically correct illustration and a thorough description. When possible, I will use layman's terms in any explanation, and when I need to use a word that you may not be familiar with, you can refer to the illustrations, glossary, and the Alphabetical List of Plants and Fungi in the back to see for yourself exactly what I'm talking about.

The natural world is simultaneously unfathomably large and impossibly small. Each member depends on the others—our lives are inextricably entwined with and dependent upon one another. History, science, and the stories of our foremothers all teach us this: the choices we make have consequences that echo far beyond our understanding, resounding into a future that we are all either blessed or doomed to share. Humans are a collective. We, unlike other parts of nature, are not driven by instinct alone; instead, we have the power to make choices. We can choose to overcome the worst parts of our impulses in the humble service of *all* life, not just to satisfy our immediate desires; or we can separate ourselves from our nonhuman kinfolk, claim absolute dominion over the land, and force it to bend to our whims. But the operative word here is *choose*. We have *choices*.

A forager must choose to consider the future. It is inevitable that if you harvest 100 percent of anything without rest or recovery and without offering anything in return, it will run dry. You may feed yourself for a day, a week, even a year, but when you are hungry again, there will be nothing left for you to take. Your belly may be concerned with the needs of today, but your mind must be communing with tomorrow. What good does it do to exhaust the present at the expense of the future?

In this way, *Gathered* is about plants and mushrooms, full of tips and tricks and time-tested information, but it is also about *you*. It is a call to curiosity and wonder, an invitation for you to place yourself in the garden of your nonhuman relations, to discover how to see and hear and understand them via the myriad ways that they communicate with us. *Gathered* is a journey toward community, interdependence, and abundance, not a guide to individual self-sufficiency.

After all, don't we all belong to one another?

We are not *on* the land, we are *in* it.

We are the forest, and the forest is us.

Gathered

Getting Started

FROM FIELD TO TABLE

THE JOURNEY FROM THE FIELD to the table is full of experimentation and tradition alike. Foragers are, as a rule, industrious and creative people who will find endless ways to streamline and improve our workflow. We repurpose tools, hoard jars and bottles and baskets, and always save paper bags from the grocery store to keep our mushrooms fresh in the refrigerator. Observing how different foragers harvest and process wild foods is an ever-fascinating experience, and when I'm lucky enough to spend time around other foragers, I'm always picking up new tricks. It all comes with experience and experimentation, and there are usually multiple ways to accomplish any given task.

Without the wealth of practical and cultural knowledge my community holds, I never would have learned that it's much easier to simply boil hickory nuts after cracking them rather than digging with a nutpick to get at the meat, nor would I have known that the best way to clean large volumes of field-dressed mushrooms is with a salad spinner. I probably wouldn't know the right way to pick nettles without getting stung, nor would I know that ground-up sour cherry pits make a beautiful seasoning that tastes almost exactly like almonds.

The work of foraging is in the minutiae leading up to and away from harvest, not in the physical act of gathering itself. It is in searching and learning how to read the landscape for signs of the plants and mushrooms you seek, in the many hours spent in the sun and rain, in learning how to see and recognize natural beings as individuals as well as in relationship to one another. The work is not in filling the basket so much as it is in emptying it again, in cleaning and trimming, in preserving and preparing what you have found.

Work is not the enemy. Work is, in many ways, the reward itself. Abundance is a gift, and having something to process means that you will have something to eat. The fruits of your own labor will always taste sweeter than the fruits of someone else's. When you forage for your food, your food becomes more valuable, more precious, more enjoyable. You will want to share it with others. You will savor every bite. Industrious labor infused with purpose and meaning that directly improves your life—work worth doing.

STAYING SAFE OUT THERE
THE BASICS OF IDENTIFICATION

SOME THINGS WE NEED TO address right off the bat have to do with very real safety concerns. Yes, *some plants and mushrooms are dangerous*! It's our responsibility to identify which is which, and to recognize the nuances of edibility where they exist.

There is no way to correctly identify a plant or mushroom's edibility or toxicity with a single metric. Very few "rules" apply equally to every situation, and you disrespect both your own body and the plant or mushroom you are harvesting by skipping over the details. In foraging, details are everything. Careful foragers may miss out on an uncertain harvest here and there, but they'll live to forage another day. The great mycologist David Arora once said that "there are old mushroom hunters, and there are bold mushroom hunters, but there are no old, bold mushroom hunters!" I think we can easily expand this to include all foragers, plant and mushroom lovers alike.

A good strategy for any identification is to start at the top of whatever you are identifying and then work your way down by studying each component individually. For a flower, that may look like starting by examining the shape, color, and number of petals, then working your way in to look at the interior parts of the flower. Is the inside a different color? Does it have long or short stamens, or stigma? Examine the color and shape of the connection between the flower and the stem, and note whether the stem has any leaves, hair, or textures. What do the leaves look like? Are they fine and lacy, or big and broad? Are they all the same shape? Do they have serrated edges, or are they smooth? Do they have lots of squiggly veins, or long straight ones? Are they arranged around the stem in a pattern you can identify, such as opposite one another or alternating? Do they appear in "tiers" like a pagoda, or does it look more random? Are they glossy or dull, thick or papery? Are they the same color on the top as they are underneath? Consider other sensory information, such as smell. The more information you can gather, the more accurate and reliable your identifications will be. One good exercise for improving your skills of observation is to bring a sketchbook into the field with you and practice drawing what you see. You don't need to be a good artist—the point isn't to make a good drawing, it is to sit with a plant or a mushroom long enough to really *see* it, to stamp its image onto your mind so that it becomes a part of your working memory.

To identify fungi, the criteria can be subtle, occasionally subjective, and very sensory-involved. Visual inspection of shape, texture, growing medium, the component parts like the cap, hymenium (underside), stipe (stem), and occasionally even excavated underground parts is a vital part of the process, but so is smelling, cutting, bruising, and even occasionally tasting. Some mushrooms need deeper examination before they can be correctly identified, which may involve bringing specimens home with you and performing various identification tests, such as spore prints (we'll get into that later). In this book, I have specifically chosen to focus on plants and fungi that are relatively distinctive in appearance, and that demonstrate enough unique identifying features to serve as valuable and relatively safe touchpoints for new foragers. You can continue building your mental encyclopedia of known species by applying your newfound observational skills to the plants and mushrooms that you encounter in the field and cross-referencing them in guides relevant to your specific area.

And by the way, just because something is technically edible does not mean that you can eat it in every situation, or that you can eat every part! You can probably think of a few commonly eaten foods that have some safe parts, and some toxic parts (like rhubarb). The ability to apply nuance and critical thinking to your identification, collection, and preparation is key to being a safe, responsible forager. Certain foods are not edible unless they are prepared a certain way, and this is often true for foraged foods. You may know that coffee needs to be fermented, roasted, and brewed before it should be enjoyed, or that many dry legumes need to be soaked and then cooked before they can be eaten. You probably wouldn't bite into a raw turkey leg unless you were *starving* at a very dodgy Renaissance faire, and you shouldn't assume that everything is edible exactly the way it comes out of the forest either. A good example of a wild food that requires special preparation is the morel. When cooked thoroughly, the morel is a highly prized, remarkably meaty fungus with a fantastic umami flavor. But if not prepared properly, you could poison yourself. Remember to go slowly and investigate exactly *what* is meant by the word *edible* in any given situation!

It's not just about separating parts either. Some foods are toxic at some growth stages and edible at others. Take the black elderberry (*Sambucus nigra*), whose berries are somewhat toxic when unripe but edible once they turn dark purple or blue. Another is the pokeberry plant (*Phytolacca americana*), an unfairly maligned native that also happens to be an important cultural food of people throughout the Southeast, particularly Black American and Indigenous people. Pokeberry shoots and young leaves are edible when prepared properly, but the alkaloid toxin *phytolacine* found in the plant increases with age, rendering the plant mostly toxic as it matures. Unfortunately, some fantastic native edibles like pokeberry often get dismissed due to widespread fears and ignorance, which puts them at greater risk for mass misidentification and unnecessary herbicidal execution.

It is an unfortunate fact that even well-known and beloved edibles may affect some people in unexpected ways (my digestive tract took a hard hit a few years back from an unfortunate experience with raw daylily shoots, one that neither I nor my toilet will soon forget). Foragers typically experiment with a larger variety of foods than most people, so we are also more prone to experiencing obscure allergic reactions and sensitivities. Start slow after a positive identification—it may not be a positive for your GI system at any stage. There is no prize for eating the largest variety of wild foods as fast as possible that I am aware of, so go slow when introducing something new to your diet.

THE FIELD
WHAT TO BRING OUT THERE

THERE ARE MANY TOOLS THAT foragers like me find helpful in the collection of wild foods. Most of them are inexpensive, and you probably have some already. Here are a few of my suggestions for gear worth investing in:

1. **Baskets.** You'll want a number of baskets eventually, preferably in a few different sizes. Small baskets are good for berries, while wide, shallow baskets are perfect for mushrooms. Deeper baskets are best for lighter items like greens and flowers. Be realistic with your basket size—overshooting may subconsciously encourage you to gather more than you can reasonably use. If I am gathering several different foods (for instance, blackberries, porcini, and chanterelles), I will occasionally take a large basket and fit several smaller baskets inside to separate my finds so I can leave one hand free. You can also separate items from one another in the same basket with little mesh grocery bags, or even delicates bags. I am partial to African market baskets (sometimes called "bolga baskets") for their durability, the ability to reform and shape them to my liking, and the fact that their bright colors mean I am less likely to lose them in the woods when I get distracted by some really good pickings. I also love backpack-style black ash baskets. These are just as durable, often locally and Indigenous-made, and can be worn in front or on your back depending on what kind of picking you're doing. These can be expensive, but they're

worth it. If you can't afford either of these options, you can do what I did for years and buy cheap used baskets from the thrift store. If you do this, avoid treated baskets—try to get unstained or un-lacquered ones to avoid contaminating the environment or your dinner should the baskets get wet (and trust me, they will).

2. **Pruners.** I have a small pair that I keep in my pocket, and they are my tool of choice for collecting twigs, flowers, grapes, sumac, and more. I think I paid a dollar for them, and by some miracle I haven't lost them yet. Pruners will prove endlessly useful for any number of tasks.

3. **Long pants, comfortable shoes, and thick socks.** You're going to ruin all your clothes if you're not careful (to be completely fair, you will ruin them even if you *are* careful), and bare legs tend to get torn up by branches, brambles, and unforeseen obstacles. In many parts of the country, urushiol-containing plants like poison ivy and poison oak are real concerns for exposed skin, and nobody likes to be munched on by mosquitoes or berry bugs. However, the most important reason for wearing long pants and socks, even in the summer, is to avoid ticks.

The CDC's best advice? Don't get bit. I take that seriously, so I look funny on purpose and tuck my pants into my socks, check myself religiously for ticks (*yes, everywhere*), and I also treat my trousers twice per year with permethrin, an EPA-recommended pesticide approved to repel ticks. Even if you don't particularly enjoy using harsh chemicals, this is a situation where I recommend it. It's better than the alternative.

4. **A good knife.** When I started foraging for things that sometimes needed to be cut or trimmed, I used to carry a paring knife from my kitchen. A multipurpose pocket knife is also a good choice. These days, I like my mushroom knife, which is curved to make it easier to cut mushrooms from trees and remove isolated areas of insect predation. It's equipped with a brush on one end for removing bits of soil

and detritus from the mushroom before taking them home. Opinel is a popular brand of mushroom knife, but they're all perfectly fine, as is a standard pocket knife, curved blade or no. The best knife is the one that can be used for a lot of things and, most importantly, feels good in your hand, because you'll reach for it often enough that you'll notice when you've left it behind.

5. **A trash bag and some rubber gloves.** Unfortunately, you're going to run across a lot of non-compostable trash that others have carelessly discarded. Get angry about it, swear and throw a fit if you need to, but don't leave it there or ignore it. Think of the environment like your house: even if someone else threw a massive party without your consent and left a bunch of trash behind, you'd still have to clean it up if you want a decent place to continue living in. The other beings that live there will thank you for it, especially the ones who don't have opposable thumbs. And, of course, I know *you* wouldn't be so disrespectful as to leave anything behind that doesn't belong there . . . right?

6. **A hori hori or a trowel.** A hori hori is a Japanese gardening tool with a few useful features: it has both a serrated side (perfect for digging and sawing) and a sharp, smooth side for slicing. It is shaped a bit like a cross between a Bowie knife and a trowel, with the benefits of both. If you're a camper, you can use it for anything from making wood curls for starting fires to digging holes for any other . . . business meetings you might need to take. If you're a lover of roots and tubers, you won't find a better tool for performing intact excavations of those high-calorie treats. A garden trowel is just fine, but you'll have to be careful not to be too heavy-handed or you may injure your dinner or surrounding plants.

7. **A cell phone.** This is controversial to some purists and Luddites who believe the only way to interact with nature is to do so without the distractions of technology, but I think it's important to have a way to call someone or find directions in case of an emergency. If you don't

want to be disturbed or distracted, switch it to airplane mode. I've been in more than one scary situation in the woods, and even just *having* a phone that I can whip out for a rescue call has convinced more than a few unsavory characters to make like a tree and *leave*.

In addition to the obvious, using a smartphone in the woods can also be a helpful way to track your finds. I used to detail all the locations and dates where I harvested or planted in a little notebook, but I was terrible about keeping track of it. Now my phone does everything for me—all I have to do is snap a photo. Saving those photos and uploading them to any of the various foraging apps can help you keep track of their locations, and can even be a valuable way to contribute to scientific research being done in your area. If you want to keep a specific location a secret, just make sure you remember to obscure it in whatever app you use!

8. **Sunblock.** It goes without saying, but foraging is often slow work and you could be outdoors for hours. Don't be the person who gets sunburnt or does long-term damage to their skin because they decided to be stubborn and go *au naturel*. Melanoma has no chill and dermatologists are expensive.

Beyond the necessities, here are a few other tools that aren't crucial, but are really, REALLY nice to have.

1. **A blickey.** This is a basket or a pail that straps to your waist and frees up both hands for berry picking. Convenient, comfortable, and easy for kids and adults to use. You can make your own with a belt and a basket, or by attaching a piece of string to a plastic bucket and hanging it around your neck.

2. **A digging fork.** Digging forks are far more efficient for unearthing the roots of large plants than a traditional shovel, and they have the bonus of aerating the soil while you're harvesting.

3. **A jeweler's loupe, also sometimes called a hand lens.** You can get these online, and they'll allow you to observe the tiniest details of nature without having to pull out a big Sherlock Holmes–style magnifying glass. Getting to know the intimate details of a plant or mushroom is crucial to building your knowledge base, and once you get your first look at the anthers of a tiny, perfect milkweed flower through your loupe, you won't be able to stop peering through it at everything you find.

4. **Waders.** These are very useful for collecting sea veggies, but my favorite use for waders is for mucking about in bogs and wetland areas. Bogs and wetlands are sensitive places, and it's generally better to take the waterways than to risk damage to the mosses, lichens, and plants that occupy them. Waders keep your clothes and shoes dry, and they'll let you access places that would otherwise require you to take a very stinky mud bath. I use them when collecting cattails and wapato, for getting to the secret blueberry and cranberry spots, and for admiring wild orchids. (I still get pretty filthy, but I like to pretend the sludge makes me rugged and handsome instead of just . . . stinky.)

THE TABLE
THE FORAGER'S KITCHEN

WHEN YOU BRING YOUR FORAGED finds home, what do you *do* with them? Often, a new forager will go out and start picking everything they can possibly identify without any forethought as to how all that food will actually be eaten. Full baskets mean future labor, and the work doesn't stop when we walk in the door. Americans waste around half of all the food we purchase, and if you aren't careful, those habits can influence your harvesting too. When we've spent time, energy, and effort tracking down and collecting foraged food, we want to make sure we have a plan for how to process, preserve, or use it in a way that honors both our labor and the food itself. The road to wilted greens, moldy berries, and soggy, bug-ridden mushrooms is paved with good intentions.

Not everything can (or should) be eaten hand-to-mouth outside! More often than not, a bad experience with a known edible wild food is due to poor or insufficient processing, not to the food itself being bad. In many instances, you must partner with the food to bring out its best qualities, which means getting to know the ingredient you're working with. Even with those foods that you *can* eat raw (I never make it out of a berry patch with an empty stomach), you'll need to find an efficient way to clean at least some of what you find, in case you have dirt or bugs that need to be evicted.

If you choose to make any of the recipes in this book, here are some of the tools you may find most useful for processing.

1. **A salad spinner.** I have a few of these in various sizes, and there is no better or faster way to clean large amounts of greens, berries, or mushrooms. Simply rinse, spin, and repeat until the water left over in the spinner is clear and free of debris.

2. **A vegetable peeler or a paring knife.** These are critical for tubers.

3. **A steam juicer.** Steam juicers allow you to extract juice from fruits and vegetables using, well, steam! The steam breaks the cell walls of your vegetation apart and releases the juice, which drips through a plastic tube into the vessel of your choosing. This is in the "nice to have" category, but it will make your life a million times easier when it comes to prepping a lot of different fruits for jelly or juice, especially fruits that have lots of seeds or pits.

4. **Several coffee and herb grinders.** These are invaluable for making mushroom and herb powders, among countless other wonderful things. Spice grinders turn dried things into powder much better than food processors and blenders do. Why do you need several?

Because they break. Constantly. I have about four or five of them and they are always in use. If I find coffee grinders at the thrift store, I usually pick them up—they run anywhere from $1 to $10.

5. **A mortar and pestle.** Opt for one made of stone, rather than wood. I use a large molcajete, a Latin American mortar and pestle with some real heft to it. You'll use it all the time, I promise. If you have the coin, you can also pick up a stone mill, which will last you forever. You can also do it the old-fashioned way by finding a heavy rock with a nice dip in it and a smaller, rounded stone that fits well in your hand. It'll take longer, but you might like the process, which one may describe as "character-building." When it comes to smashing things, I've offered a few easy-to-obtain alternatives in my recipes.

6. **A food processor or a good blender.** I use a low-end Vitamix these days, but having a tough blender really does make a huge difference when you're processing larger quantities of foods that you want to get extra smooth or fine. I like the fact that I can make nut flour, silky-smooth vegan cheese, and prepare larger volumes of food than a spice grinder will allow.

7. **Sharp knives.** You don't want to mess with dull knives. A lot of the foods we work with can be tough at first, and using a dull knife is a recipe for an unwanted emergency room bill (ask me how I know). It's nice to have a few different knives for different tasks, but don't go overboard—a chef's knife, a vegetable cleaver, a serrated knife, and a paring knife will offer plenty of blade coverage for most foragers.

8. **A French press (or a few).** Nothing is better for separating plant matter from a liquid, especially when you're making syrups and teas. They are also useful for fermenting, since they make it easy to depress flowers and fruits below the level of the liquid for maximum infusion. I use my French presses all the time during flower season, and it isn't unusual for me to be running six or seven infusions at once.

9. A food mill. These are great for berries and soft fruits. I use an OXO food mill, but all brands generally work the same way, using a crank to press out fleshy pulp while separating it from the seeds and skins using a strainer. Cleaning them can be a chore, which you can only fully appreciate after you've spent an hour trying to push blackberry seeds out of the stupid little holes with a toothbrush, but the time you'll save by using one is well worth the annoyance.

10. A dehydrator. More on this later, but a dehydrator will significantly level up your foraging game. Watch pounds of mushrooms turn into dry ounces in mere hours, ready for powders and long-term storage; make your own foraged fruit leathers; and see armfuls of greens reduced to quart-sized jars. There are many other ways to dehydrate depending on your food quantity, climate, and space limitations, but my sturdy little machine can be heard humming away at least a few days a week during busy seasons. A *lot* of the recipes in this book are made easier with the use of a dehydrator.

11. A nutcracker. I have a few different nutcrackers that I use for different nuts. For softer nuts like acorns, I prefer a smaller nutcracker, usually a screw-style that gives me a bit more control and versatility rather than a crank, but when you really get into the nut crackin' life, you might want to invest in something more heavy-duty, usually something made of cast iron with a hopper that can handle more nuts at once. Davebilt is a popular choice for acorns and hickory nuts. Harder nuts like black walnuts require extra force, and traditional nutcrackers will be unable to penetrate them. You can smash them individually with a hammer, a stone, or your Kia Forte, or you can invest in a nutcracker specifically built for black walnuts like Grandpa's Goody Getter. Cracking nuts is always better with good company, so do yourself a favor and convince everyone you know to invest in one kind of nutcracker, then have them bring all their nuts over and turn it into a social activity.

12. A digital scale. These are necessary for lacto-fermentation, and have a number of other uses besides. When making a recipe, most of us are accustomed to being able to look at the bag, box, or can that our food comes in to find out how many ounces or pounds it is, but when foraging, we don't have this information.

Most of these tools are *optional*, and you don't need to go out and start buying a lot of things all at once. The vast majority of my tools were either thrifted or gifted, and I have accumulated them over the course of many years. Determine what you like and what is worth investing in before spending any money on gadgets you may not end up using. If you don't use a lot of nuts or grains, you might not need a mill or a fancy nutcracker. If you're not partial to using or making juices, no need to buy a juicer.

Foraging is better when practiced in community with others, and it tends to be cheaper too. Resist the urge to go out and buy everything you might possibly want to use, and instead, work on nurturing a community of like-minded, trustworthy people who can spread out bigger purchases among themselves depending on the individual's specific needs or interests. Treat your tools like a library, if you can. If one of you has a big, fancy dehydrator, maybe you don't *all* need to go buy one. You can only crack one type of nut at a time, so maybe you don't personally need one of each type of nutcracker. Figure out what foods you love working with most, and then invest in tools that let you do that work more efficiently.

THE FUTURE
PRESERVING WILD FOOD

FORAGING SEASONS TEND TO BE brief, some as short as a few days, so very quickly you will learn to prioritize collecting for future use as well, should the available supply allow. Many of the recipes in this book will suggest using preserved finds from previous seasons as you build your larder. There are a number of different ways that we can preserve what we find, and here are a few of the methods I discuss in this book.

Fermenting. This is probably one of the best (and tastiest) ways to preserve many foraged finds. Whether you're making vinegar, kimchi, cordial, wine, kraut, kombucha, or some other crazy ferment, they tend to be showstopper flavor bombs that elevate your cooking way beyond "meat and potatoes" and into the realm of the gourmand. You don't need much to get started, but you *will* need to make sure you can tend to your ferments regularly so they don't explode like the ill-fated plum wine I gave to my forager friend Alexis Nelson way back in 2020, which I later learned managed to hit her ceiling with significant force. She very graciously told me that it still tasted great.

Dehydrating. Dehydrating is a fantastic way to preserve mushrooms, herbs, alliums, fruits, greens, and florals for powders and teas and more. It can also keep certain dry foods from spoiling, and can intensify the flavors of ingredients like mushrooms, alliums, and spices. Running a dehydrator can be costly energy-wise, but you can also build an herb rack, string-dry, strip-dry, or sun-dry many different ingredients for future use. I do a lot of my summer dehydrating on the dashboard of my car, which costs me nothing and works like a dream. It is worth keeping in mind that the initial energy cost of dehydrating tends to be mitigated by the overall cost of leaving big bags of mushrooms or greens in the freezer for months or years at a time, and can certainly help reduce waste in the long run. I like dehydrating my veggie scraps, powdering them, and adding them to my meals for an added hit of nutrients. To keep dehydrated food fresh in long-term storage, invest in desiccant packets. If you don't want the packets touching the food, tape them to the underside of your jar's lid.

Canning. There are two typical ways to can food: water bath canning and pressure canning. It can be difficult to find proven canning times and recipes for certain foraged ingredients, but in general obvious substitutions are okay to make (for instance, using canning times for spinach to figure out how long you should can nettles). Pressure canners reach much higher temperatures than the normal boiling point of water, and those higher-than-boiling temperatures are necessary for eliminating certain bacteria that may be present. One such bacteria is *Clostridium botulinum*, which is responsible for causing a neurotoxic condition called botulism in high pH, anaerobic (sealed) environments. The acidity of our ingredients is a key factor in determining what can be safely canned in a water bath versus what must be canned in a pressure canner. Highly acidic foods such as berry jams and meatless pasta sauce can generally be canned safely in a water bath, but mushrooms, low-acid vegetables, and meat must always be pressure canned. If in doubt, buy a pH testing device and use it to make sure your food is acidic enough before water bath canning. The canning time and exact pressure will be determined by the ingredient, the size of the jar, and your barometric pressure. As a backup, you can always freeze or eat fresh.

Freezing. Some ingredients, particularly certain mushrooms like edible polypores, are best preserved by freezing. Dehydrating can damage the texture of some species of mushrooms, most notably chanterelles; therefore, it makes sense to avoid dehydrating those unless you plan on powdering them and adding them to other foods or spice blends later on. When freezing, try to avoid doing so in oil or butter, as this can make freezer burn flavors more pronounced.

Alcohol extractions. Alcohol (or glycerin, if you're a teetotaler) can be used to preserve flavors, and sometimes to create more. Certain flavors are more stable when their volatile compounds are preserved in alcohol, such as vanilla, and in many cases, they can be used to flavor liquors and create bitters. In addition, like vanilla, alcohol extractions can be used to flavor candy, baked goods, ice cream, and more. Alcohol extractions are often the first step when creating tinctures, and while I don't touch on tinctures in this book, they can be a good way to preserve medicinal plants.

If I can offer a word of encouragement—don't be too stingy with what you make! Use it. Enjoy it. Share it. You worked hard to find, gather, and process it. Nature is abundant, full of gifts in every season, and if you don't eat your way through what it offers you, you're going to run out of room for new treasures.

The Stillness

White is the ground, white is the mist,
What enchanted quiet land is this?
Under what heaven do I walk?

Jacqueline Elisabeth van der Waals,
untitled

JANUARY 2020
Quincy, Indiana

My hands are frozen and the saliva behind my right cheek tastes bitter. It doesn't seem to be enough that I lost my home and am living out of a beat-up RV that I've named "Kerouac," a tongue-in-cheek reference to the fact that it is not remotely safe enough to be "on the road." No, it couldn't be enough that I am going through a terrible divorce, or that said divorce was initiated a *week* after my beloved cat tragically passed away, or that my *other* cat died unexpectedly of a heart attack just a few months later, leaving me alone and wondering what I'd done so wrong to be left behind by the beings on earth I loved the most.

Nope, not enough at all; my pipes just *had* to freeze on a Friday. I check the faucet again hoping for a different result—nothing.

Swearing under my breath, I shove my hands under my shirt and press them to my stomach to warm them, a trick I learned during a one-off horseback riding lesson in my early teens. It works, like it always does, but my core tenses unpleasantly at the sensation. I pull on a beanie, hoodie, heavy coat, snow pants, woolen socks, boots, gloves. Coffee? Right, I remember, no water. I briefly consider boiling snow on my single induction burner, but somehow that feels like a bridge too far.

My best friend, Pascale, owns the farm where I've parked my RV, and lives in the house a little farther down the driveway. It was a foreclosure when she purchased it, practically condemnable upon arrival. We toiled day and night to make it acceptably habitable, hauling dumpsters worth of filth from every nook and cranny, tearing up maggoty carpet, laying tile, fixing a multitude of bewildering hazards, and even wrangling basement snakes (yes, that's a thing). I took out a bank loan, bought that rickety fifth wheel online for the grand sum of $2,500, parked it in the field beside the driveway, and moved in. Exhaustion, survival mode. Everything beyond a roof and a place to sleep is just details.

I can figure out details later.

Despite being the one to convince me to move to the property in the first place, these days Pascale regularly urges me to come sleep in the house for the winter. "You're going to freeze to death one of these nights, I'll have to use a crowbar to get you unfrozen from the bed."

I appreciate her concern, but refuse her pleas every time. "I need to make this work. I have to prove to myself that I can do it."

There are lots of things that nobody tells you about "van life." Your pipes will freeze. The roof will leak. Fungus will grow in the carpet (thanks to the roof, which again, *will* leak). The propane tank will run out at the worst possible time. If you try to charge your phone and use the microwave at the same time, you'll trip the breaker. You will spend far too much time knee-deep in problems that involve your own literal shit.

Screw it.

Time to go, walking will warm me.

I flinch in anticipation of frigid air as I open my flimsy metal door, but it never comes. I am met with a white sky that gives no indication of the hour, six perfectly level inches of snow on my steps, and giant, fluffy flakes gliding down without so much as a breath to steer them off their course. My heart slows automatically. The world is almost uncomfortably quiet. Of course.

The Stillness has arrived.

The Stillness is a time of the year when nature's silence is punctuated only by the howling voice of the wind playing percussion with austere branches, the crunch of snow beneath boots, and the rhythmic scraping of ice from frozen windshields. However, in certain meaningful moments of sheer perfection, when the clouds are heavy with oversized flakes of snow that fall straight down in a perfect line, and time has lost all sense of certainty, the Stillness rewards careful observers with a particular peace. The rules of time and its analog, sound, are disrupted beyond recognition.

Music is many things, but primarily (or perhaps more usefully) it is an art form that takes up space that is not physical, but temporal. Music is so much more than the shaping of sound, because sound is so much more than vibrations hitting our eardrums. Shaping and producing organized sound allows us to explore not only pitch, rhythm, and timbre, but our very perception of *time* as it exists. John Cage, composer and mycologist extraordinaire (as luck would have it), managed to demonstrate this through his iconic *4'33"*—a piece comprising four minutes and thirty-three seconds of total inactivity. He just sat in silence, observable. Some might call Cage's work performance art, rather than music, but even the biggest stickler for the rules has to admit that *4'33"* is anything but silent. Nobody can hold their breath for four minutes and thirty-three seconds (to be fair, I'm sure *someone* can, but certainly not me). Most are unable to avoid shuffling, sneezing, or even unconsciously tuning into the hum of the radiator to regain their sense of time within the context and comfort of sound. Even the body's systems can't help but make noise—your pulse will start to play a soundtrack in your ear if your environment is too quiet.

A similar aural discomfort exists when one is outdoors amid a heavy snowfall. The sound of the Stillness is discrete and singular, calming and unnerving at the same time. It marks a time and a place, a particular weather pattern, a specific melody within nature. There's a reason for this. Snowflakes are highly porous owing to their classic six-sided shape; therefore, even a few inches of accumulated snow create a series of sound traps. The effect is the illusion of silence, and perhaps even the illusion of stillness. Billions and trillions of individual actions take place during snowfall, as each perfectly unique flake sacrifices their microscopic body to form a massive terrestrial blanket, insulating the fragile plants, fungi, and creatures slumbering below.

When snowflakes melt and lose their distinctive shapes, either turning to water or solidifying once again into ice, the spaces created by their lacy edges dissolve and their ability to absorb sound is reversed, amplifying instead of ingesting.

The Stillness is fleeting, its magic only appreciable during the soft moments when it is briefly recognized, the moments that hang, like a flake on an ungloved hand. The world is awash with sound, even when human voices and instruments are quieted. This was perhaps the defining goal of Cage's work—to remind us that sound does not exist within a hierarchy; that noise is simply sound we believe to be ugly or out of place, that silence is a necessary complement to sound, if true silence even exists at all. Stillness may not be synonymous with silence, but if anything like silence actually exists, it is the quietude that sets in with a heavy, windless snowfall.

I'VE DRIVEN AN hour and a half north to reach this wildlife area in northern Indiana (a foolish decision, given the state of my tires), and as I shift my rusty 1991 Honda Accord into park, I take a deep breath. The weather has changed—I can see the snow streaking almost sideways as it dances past my windshield. The icy wind smacks me as soon as I open the door, painful on my fingers and cheeks, but invigorating in its intensity.

I enjoy this transient weather for what it is, but impatience for spring pricks uncomfortably at my fingertips. I take some comfort in observing the fattening terminal buds on the leafless trees. Despite their thin casing of clear ice, they are swollen with promise. Refocusing, I remember why I am here. Today, I am looking for something on the ground. If nothing else, it should be easy to spot, given its bright red color. But first, I need to find some oaks.

In my mind's eye, I am eight years old again, nearly nine, wearing a red winter coat two sizes too big for me, smeared with soil and sticky with pine sap. I walked with my friends, all of us the same age, a ragtag posse of feral homeschooled children, generally happiest when shoeless, scraped, and filthy from head to toe. Our exhausted hike leader prodded us up the side of a Pennsylvania hill, to a summit locally known as The Knob, where the whole town of Milford could be seen

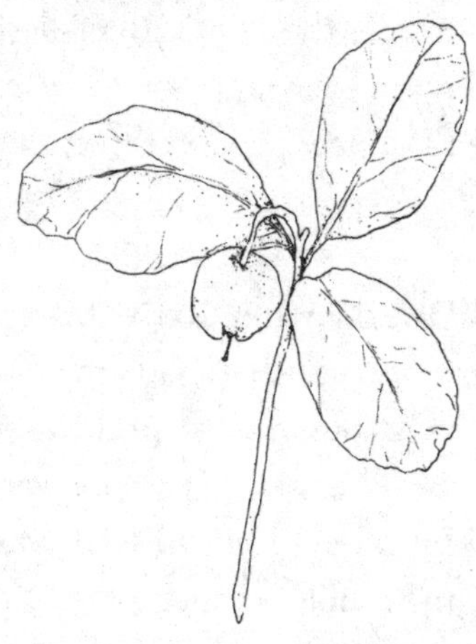

Gaultheria procumbens, Wintergreen, American Teaberry

stretched out below us, three states visible from one locale, cleft once and then again by the winding Delaware River.

We were less concerned with a great view than with crushing in a particularly good ice puddle under our boots, collecting Really Good sticks, arguing about the things children find important (like which one of the Pevensie children each of us would be if we were characters in *The Chronicles of Narnia*), and making altogether too much noise for the relative peace of the forest. Turning from the fray, I spotted my friend Linnea bending over and plucking something from the ground. Her short, boyish hair framed her serious face, and I could see her raise something to her lips and watch it disappear into her mouth. "What's that?" I asked quietly, not wanting to alert our warden in case she had done something horribly wrong. "Wintergreen berries," she said, in a matter-of-fact voice I had come to know well. "Try one." She pointed to a small patch of short plants with friendly looking leaves, far too awake for the frost, adorned with bright red berries. "They're good. They taste like mints, but better."

I hesitated, but my curiosity won me over. The ominous scarlet of its outer flesh worried me, as I had been warned that red berries were often poisonous, but if I was going to die a horrible death, at least I wouldn't be the only one in trouble. With a furtive glance, I snuck the berry into my mouth. Huh. It really *did* taste like gum, a bit mealy inside, the white inner flesh reminding me of a too-old apple, but not unpleasant. *Surely something that tastes good can't be dangerous,* I reasoned, picking three, then five, then ten more.

OUR LEADER GLANCED over, did a double take, and barked "*STOP*" in a voice that made my blood run cold. Something I thought to be rage (but in retrospect was more likely terror) boiled behind her eyes, and she asked, "*What have you done?*" I froze and dropped my handful of red treasures—I was always the coward, terrified of getting in trouble. Linnea piped up, honest as always, arguing "they're just wintergreen berries. They're not poisonous, watch," as she popped another one in her mouth.

Banana Confidence

Out of an abundance of concern for the safety of any curious readers, I should point out my childish risk-taking isn't an endorsement for identification experimentation. Plenty of delicious things are toxic (if they weren't, nobody would ever get poisoned)! Whether you are a beginner or a seasoned forager, it is imperative to be *absolutely, 100 percent doubt-free* in your identifications. My friend Sam Thayer calls it "banana confidence," referring to the level of certainty that you would have if you were asked to identify a banana. And almost 100 percent of the time we could get that right—we know bananas really well. To get to that level of banana confidence while identifying wild foods, you will have to observe them in their environments for quite some time before you ever eat them.

"You don't know that they're not poisonous. Spit that out right now and I don't want another person putting a single thing from this trail in their mouths—I mean it," she growled. "Keep moving."

My heart pounded with red berry–hot embarrassment as I wished with all my might to sink beneath the forest floor, never to be remembered again, watching my friends walk dutifully down the path toward our destination. Linnea hung back with me, observing my discomfort with mild amusement. She always had a knack for recognizing when grown-ups were being foolish, and a bewildering habit of remaining entirely nonplussed in the face of their irrationality. Ensuring that the coast was clear, she opened her hand, still full of berries, and popped one in her mouth. "I know what they are, and they taste good. Wintergreen berries taste like wintergreen. I don't know why she's so mad. And they *are* food."

A CIRCLE OF oaks is ahead, a mighty battalion gracing the steep slope leading down to a creek. To my good fortune, the snow is not quite so deep here (thank you, gravity), and I catch a glimpse of a red berry poking out beneath the snow, shrouded in three familiar, elliptic green leaves. I pop it in my mouth and feel the light, minty crispness pass over my tongue. It seems that wintergreen does, indeed, still taste like wintergreen.

I sigh. My old friend, you are just as I first found you so many years ago.

Have I changed since last we met, or am I still the same?

This wintergreen *Gaultheria procumbens* is not a true mint; rather, it is a member of the heath family *Ericaceae*, which makes it a distant relative of the blueberry. You have probably tasted it, especially if you frequent novelty candy shops and happened to try a gum called Teaberry (one of the many common names for wintergreen). It has a long history of use as a flavoring, including, of course, in mints, gum, toothpaste, even chewing tobacco. But wintergreen also happens to be a surprise ingredient in a beloved beverage you may know as root beer.

Finding food in the forest is addictive. Once you find your first plant, your first mushroom, you'll be chasing that high for the rest of your life. Even when you think you've gotten to know the ins and outs of how to work with an ingredient like wintergreen, you can always learn or dream up something new. Chop and infuse the leaves in hot water and they'll turn the water ruby red, do the same thing in alcohol for a few weeks and it'll go emerald green. The applications are only limited by your imagination. Tea, ice cream, bath scrubs, candy, Christmas cookies, the list goes on.

I am a creative person, and much of my inclination to create things is inspired by my deep sentimentality about nature. At this point I don't think I could look at a plant as anything other than a sort of person. When you return to the same places year after year, their faces are those of old friends, warm and inviting, providing a consistency and familiarity that not even people can match, and each year I find myself sitting with the children of the annual green plants I visited with the previous year, to catch up with old trees that have aged another year like me, and to meet the new fruit of my fungal friends. There was once a time when their spirits were honored regularly with intentional human company, despite less favorable conditions, and it is my hope that one day more people will provide them with this essential kindness.

This is easier in summer, when everything is awake, but I try to make a point to visit my friends in winter as well. I know how lonely I get as the days begin to darken.

With the familiar feeling of something knocking around in my creel once again, I set off through the snow. We're not done yet, and there is still so much to see. This time, I am in search of elm.

Elm trees are a mushroom forager's best friend, whether they know it or not. Elms are what we call *indicator species*, meaning that their presence frequently coincides with the presence of a plant or mushroom of particular interest. Elms are preferred hosts for many different species of edible fungi, most famously morels for a few short weeks in the spring, but during the winter, dead elm trees shelter fragile enoki mushrooms, *Flammulina velutipes*, beneath broad sheets of sloughing bark. After a few frosts, you might be lucky enough to find a bouquet of squat mushrooms, sticky caps like honey-glazed, golden dinner rolls, stems fine and downy, hence their common name "velvet foot."

Spore Printing

The color of a mushroom's gills doesn't always match the color of the microscopic spores that they shed, creating an opportunity for a rather fun identification activity. Cut off the stem and place the cap, gill-side down, on a sheet of paper, then cover it and wait a few hours before removing the mushroom to reveal a perfect impression of the gills made from the dust of spores. It is customary to make two spore prints at once, one on dark paper and one on light, to make sure at least one will be visible, but I prefer to make just one on a neutral or reflective surface like glass or aluminum foil. Often, intentional spore printing is an unnecessary step in identifying enoki, as the upper levels will snow generous helpings of white spores on the mushrooms beneath them, creating a sort of natural spore print. However, it's never a bad idea to double check your identifications before chowing down if you're not sure.

To identify enoki for the first time, you will need to "take a look under the hood," so to speak. Turn it over in your hand and examine the cream-colored gills, like a fan with each fold separated from the other by little gaps that let the sunny color from above peek through, its shorter gills extending from the outer edges of the margins.

I have sat through many grumpy lectures where fellow mycophagists ho-hum the culinary virtues of the humble enoki, complaining of insubstantial yields and toilsome cleaning duties, arguing that this somehow proves they are far more trouble than they are worth. I disagree. Half the joy is in the preparation. That's where I spend time with an ingredient, getting to know its little quirks and working with it to showcase it at its fullest potential. And with enoki, the potential is quite vast. Their subtly mucilaginous texture has a thickening effect on a broth that can be quite pleasant, and on the other end of the spectrum, when pan-fried and finished with sesame oil and miso, they can add a delightful umami crunch in place of fried onions.

I suppose one man's trash is another woman's treasure.

My left boot is getting wet through the spot where the heel has come away from the spur ridge. I should stuff a few plastic bags inside to insulate it, but I only remember to do that after it's too late. I find myself wondering if I could learn to repair shoes. I'm reminded of some of my favorite fantasy books by Terry Pratchett, where he introduces the "Sam Vimes theory of economic unfairness," also called "the boots index," which postulates that wealthy people remain wealthy because they are able to afford to buy nice things which last for years, while poor people end up stuck in a cycle of purchasing cheap goods because they can't afford the nicer ones, which means they wear out faster and new cheap things have to be bought over and over again to replace the old cheap things. After a while, the rich man still has a good sturdy pair of boots after having spent a single sum one time, while the poor man has a pile of wrecked shoes, having spent a little bit of money ten times over.

The boots, the endless problems that plague Kerouac, the failed marriage that continues to embarrass and slice away at pieces of me, like a shawarma being shaved away one slice at a time. My life is full of things damaged beyond repair, and I worry that I may be one of them. Worse, I fear that I'll just end up replacing everything broken in my life with the same cheap crap over and over out of the same desperation until I've wasted my entire life wearing ill-fitting boots with holes in them for no good reason.

Just a little farther now. I remember the tree I have in mind, the standing dead elm. I picture where I found them last year. I've always had a good memory for landmarks, despite the fact that I can't navigate my way out of a paper bag under normal circumstances.

Down there, in the ravine, I spot it. The tree is tall and craggy, huge limbs littering the ground beneath the snow, evidence of storms past. Every year, it loses a little more of its crown to the forces of entropy, and more of the tree's decay is exposed to the elements through long fissures in the sloughing bark.

Part of what I have always found so wonderful about fungi is that they are crucible and catalyst for the circle of life, truly a beginning and an end all at once. Fungi break down the things that can no longer grow, and in doing so, they infuse them with new life. The death of a single tree sets off a chain reaction that results in fungi moving in to break down lignin (the tough polymers in trees that provide them with structural integrity) and making space for moss and lichen to move in, cooking rich humus underneath.

Have you ever dug into a rotting log in the summer and lifted the damp, velvety, red-black wood to your nose? It has a certain smell, a rich and sensual petrichor, evidence of a compound called *geosmin* caused by the busy machinations of microscopic *Streptomycete* bacteria, and you can probably see tiny threads of white mycelium weaving through if you sift very carefully. Sit with the log a bit and you'll notice tiny mushrooms popping through the thick blankets of moss, crusty shield lichens forming entire armies on bare areas, insects crawling through the vegetation and burrowing homes for themselves in the softened wood, and hundreds upon thousands of unexpected sparks of life colonizing every available nook and cranny. And yet, miraculously, even *that* is not the end of the log's life. When its new inhabitants have spent every last gift of energy the log has to give, it will break down into rich soil, and perhaps a new tree will grow in its place, as has happened again and again for millennia. (The Log Lady in *Twin Peaks* may have been onto something when she remarked that "one day, my log will have something to say about this.")

I forget my physical discomfort for the moment, beholding the tree and lifting the bark slightly. I see them huddled, waiting; a massive cluster of slightly flattened mushrooms, golden brown and sticky, pressed between the bark and the underbark. Removing my gloves, I peel the bark back gently and click my knife open, removing the largest and cleanest specimens and placing them in my creel. Within a few minutes the newly exposed enoki are dusted in snow. It seems wholly unlikely to find a mushroom that so closely resembles fresh egg-washed bread hiding in the most desolate and lifeless of places.

I thank the tree with a familial pat on the trunk, press the bark back in place, and promise to return in spring before setting back to find the trail again.

Crunch.

I feel in my pocket for my cell phone, then remember I left it tucked in my glove box along with my wallet. I don't know what time it is, and my hands would be too frozen to use it anyway, so I wade back in the direction of home. My mouth feels dry and tacky, and the snow is making it difficult to walk. After a few labored steps, I slip and slide all the way back to where I started. Why did I go all the way down the slope? *Maybe there's a better way to climb back up this thing,* I muse, sweeping the landscape for gentler terrain or at least a few handholds, but there is nothing obvious to be found.

It isn't the first time I've gotten stuck from utter carelessness, nor will it be the last. Still, it's incredibly humbling to realize when you've made a foolish mistake and paid too little attention to the fact that if you are going to *go* somewhere, you also have to find a way to get *back*.

In my first year of college, I decided to go camping with my then-boyfriend in the middle of November, which wasn't strictly permissible at my school, so we went deep into the woods in a nearby park to find a place to set up our sleeping bags. (In case you were hoping this was going to turn into *that* sort of book, I'm very sorry to disappoint you, but *you* try getting busy when it's 12 degrees before the windchill.) The most secluded place we could find within walking distance was at the bottom of a deep ravine in Playwicki Park outside Philadelphia, about a hundred-foot climb at a 70-degree incline, with in-use train tracks at the top. (Literally *none* of this was a good idea.)

The spot we selected was hard to access by design, thick with brush, as post-adolescent paranoia had us convinced that someone from the university would find us and turn us in. Getting down there was one thing, but getting back out? After two inches of snow had fallen during the night? *Oof.* We emerged, eventually, but it took over an hour, I fell twice, we were both filthy, and it took me hours to get all the twigs out of my hair.

And yet, here I am, stuck in a ravine again.

Am I doomed to repeat the same foolish mistakes over and over again? Will I *ever* learn?

JANUARY 2018

Indianapolis, Indiana

I walk every night, rain, shine, or snow. I follow the back alley to Broad Ripple Village, where I play a game of pinball at the arcade. I make a game of retracing my boot prints on the way home, making them perfectly equal going and coming, a heelprint on each end.

I don't have to walk at night, and perhaps I shouldn't, but I choose to anyway. The streetlights reflecting off the snow-bent pines makes me feel as though I am walking through a tunnel to some magical world.

Every Wednesday I go to an open mic night at the local pub, Books

& Brews. My sanctuary, my church. I play the owner's keyboard and sing covers of mournful tunes by sad Irish men for my fellow congregants. We are made up of twenty- and thirty-something amateurs with guitars and a limited repertoire, senior citizens who've played the open mic circuit for decades, the occasional singer-songwriter with a rare talent who inevitably moves on from Indiana. Very few women play. Some come to sing but disappear after a week or two.

Between plates of fries, pints of beer and cider, and our three-song set limit, we form a camaraderie; a collective of people with nothing better to do on a Wednesday night. Some of us pair up and perform duets, joining in on each other's favorites. I sing harmony, filling out the songs I hear week after week.

Most of us are just looking for somewhere to belong for a few hours.

My friends, a group of men in their twenties, bring dates to show off their talents, occasionally turning those dates into girlfriends and fiancées.

I invite him a few times to come watch me play, or to bring his guitar and play with me like we used to, but he declines. It's late, he has class the next day, he's supposed to play a game online with friends.

No problem, I hear myself say each time. *No big deal.*

I stop asking after a while.

When it comes time for my set, I pick a song from the days when our long-abandoned duo, The Bird and Brielle, used to play together in places like this.

I try to remember the melody he sang over the lyrics, but catch myself slipping into the more familiar alto line I used to sing.

Been talkin' about the way things change . . .

JANUARY 2020

Still Lost in a Ravine

The sides are too steep to climb, and nobody is here to help steady me if I lose my grip.

Well, this time I'll have to climb out on my own, but at least there is no train waiting to pancake me at the top. I brace myself, and start walking along the bottom of the ridge to find a way back up. The best thing in this situation would be to find a dense patch of saplings or perhaps some stable-looking stones against which I might be able to leverage myself, but the trees are too far apart and the snow is too deep to see anything that might serve as a hand- or foothold.

Crunch.

I've walked for at least half a mile, probably longer, and the top of the ridge seems farther away than it did from where I started, because of *course* it does. Clearly, I chose the wrong direction. Perhaps if I'd gone the other way, the topography would have been more in my favor. For the first time, my irritation turns to concern. Nobody knows where I am, and frankly, "nobody" also includes me, and at this point my frustration and discomfort turns to a very real fear.

To my left, a widening stream flows through the landscape, carving languid curves in the drifting snow on either side. Flanking me on both sides, the cavern grows ever more impossible to traverse, almost as if it is rising out of the earth.

My own foolishness is closing in around me. I have no lifeline, inappropriate gear for the weather and the task, and worst of all, *nobody knows I'm here.*

I'm alone in the Stillness, but even if it *feels* like time is spinning out without really passing like a bald tire in the snow, it is increasingly becoming my enemy. The longer I stay out here, the more perilous my situation will become.

It's easy to forget how capable any one of us is of making dumb choices like this when we stop paying attention. Reading a text while driving, leaving a candle burning, deep-frying a frozen turkey—most of these are *incredibly* dangerous things people do every day because they haven't internalized the risks. And you *have* to know what might be waiting for you on the other side of a careless choice, because nature is neither cruel nor kind, she simply *is*.

There is no safety net, no magic savior, no mystical wolf family willing to take you in, clothe you in furs, and raise you as one of their own.

I make more mistakes than I'd like to admit to.

But I will admit to them. I *learn* from my mistakes.

I can do this, I think, ignoring the panic that threatens to claw its way up my chest. *I solve problems. There has to be a way up . . .*

Wait. I stop for a moment. *I'm doing this wrong. I need to take a closer look at my surroundings.*

Immediately, there is nothing, just as there was before. But in the distance, back behind me, I realize that I can hear something.

A faint hum, like a distant crowd.

The waterfall. I remember now. A main feature of this park is a waterfall, and the waterfall has wooden stairs leading up and down to a lookout point. If I can reach the stairs, I can get back up.

I don't quite internalize this fact, but looking back on this experience I will realize that I am racing against time. I don't feel the cold anymore, especially not in my feet, which are soaked through and only getting colder. My right hand is curled around the strap of my creel as though made of ice itself.

Following my own footprints, I walk back in the direction that I came.

SPLASH!

My heart sinks as my boot connects with a thin sheet of ice under the snow and pushes through, not stopping until the frigid water hits my knee.

Why are you such a disaster? Voices begin to hiss in my head. *You're like a child, always making a mess, always forgetting your head. My life is better without you. More peaceful without you. More structured without you. More fun without you.*

I limp onward, keeping as far from the edge of the creek bed as possible, swearing incoherently to and at myself, wishing I had gone to the movies or perhaps foraged some steamed dumplings or a pizza instead.

Oddly enough, I notice long, green grass spikes can be seen poking up through the snow in bunches, at intervals. Upon closer examination, I am baffled to see field garlic, *Allium vineale*, defiantly braving the snow. The sight of something green and growing is welcome, but *weird*. Even though I know in my head that this plant doesn't mind winter, it still feels strange to see it thriving in such harsh conditions.

Allium vineale is a weed in North America, if you subscribe to the idea that a weed is a plant growing somewhere that it isn't wanted. It goes by dozens of names: crow garlic, wild onion, yard scallions, and more, and is enormously resilient. It is a member of the plant family *Liliaceae*, which contains not only lilies, but also onions and garlic. Plants like onions and garlic that produce bulbs are ingenious in their self-defense strategies—building up layers of modified leaves, thick and dense, storing enough energy to sustain the plant even through the most bitter conditions. *Allium vineale* is considered an invasive plant, thriving just about anywhere it finds itself and occasionally pushing aside more delicate native plants to do so.

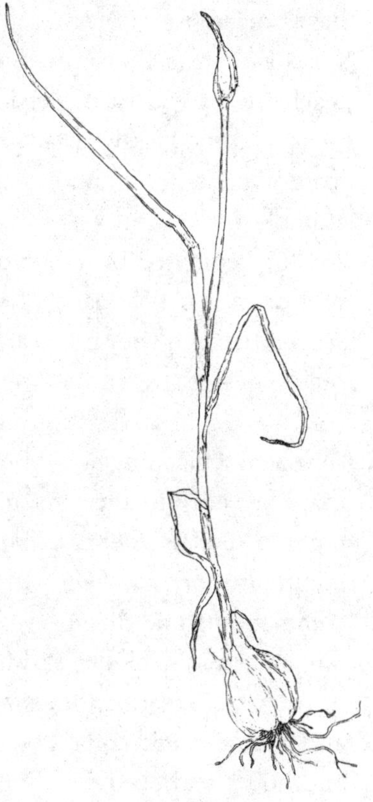

***Allium vineale*,**
Crow Garlic, Field Garlic

I am grateful for this plant today, and I might even relate to it. Its stiff determination to continue growing even as the snow piles heavier and heavier on top of its head gives me a spark of courage that I didn't know I needed quite so badly. I kneel down and push the snow aside from atop a small clump, ignoring the cold and the wet as I collect a fistful of slender tube-shaped leaves.

What seems like hours pass as I retrace my steps, though realistically it was probably only twenty or thirty minutes, but finally the whisper becomes a full-blown roar. I sense the mist peppering my face, though I cannot feel it, I come around a bend, and there it is, in all its post-glacial glory. Miraculously, the imagined stairs that I had bet my frostbitten toes on are, in fact, present and accounted for. They even look relatively safe despite being quite wet, as the waterfall hasn't begun to ice over yet.

I steady myself against the railing, hoist myself up a few feet to reach the base deck of the lookout, and begin to climb. It is hard work, and the glassy surface of the steps proves slicker than I anticipated. It takes two hands just to keep from falling, and even then I stumble.

I am reminded of the winter I spent in Olafsfjordur, Iceland, as part of an artist residency. I remember the penetrating cold, how it seeped beneath your pores and entered your bloodstream, how your organs seemed to run slower, your body weighed more, a rending cold that did its job so thoroughly that it eventually stripped the concept of any meaning at all. But it wasn't just the cold, it was the *dark* that chewed into the very marrow of your sanity, sunrises bleeding into sunsets, the mocking sun never actually ascending over the mountains for a moment of thaw. There is something brutal about it, commanding and direct—*you must rest.* You try to fight it, to maintain a normal schedule, and find that your body simply can't handle it. You work, creating things with your hands, painting and sculpting and baking to add color to your days, body untethered from minutes and hours and even days. You eat when you're hungry, sleep when you're tired, doing everything you can to make your isolation bearable, creating a cozy fortress to wait out the inhospitable reality of what lies out there. One is warmer for the knowing, glowing with a heat that can only be made through the stark contrast of relentless cold. And perhaps, when the nights are clear, you can step outside and witness the aurora, ribbons of light dancing so slowly that blinking is the only way you can prove that they are moving at all. Time working outside the custom of my expectations again in winter, weaving lights on a loom too large for me to imagine.

The aurora borealis is one of nature's many light shows, visible only when solar disturbances tug on Earth's magnetic field (the poles), and most apparent in winter when conditions are darkest. According to astronomers, turbulence on the sun results in cosmic ripples, known as Alfvén waves, which send supercharged electrons whizzing into Earth's upper atmosphere. The particles are flung into the ionosphere, the liminality between Earth's breathable lower atmosphere and the vacuum of space located fifty to five hundred miles above, where they collide with neutral molecules they find there. The resultant energy surge from the molecular crash-landing creates the world's slowest fireworks show. Oxygen molecules are responsible for the eerie greenish yellow and red colors we can witness from Earth, nitrogen for the blue.

What nobody tells you about waiting for the sun is how it feels when she finally lifts her head above the mountains.

The sun touches my face for the first time in two long months, and I melt. *Hello, old friend.* I touch my face. Are those tears? I can feel the sunshine burning away weeks of frost that I didn't know was living on me.

How could I have forgotten how much I missed her?

At the top I am greeted by a freshly shoveled path, a sandwich board sign warning trail users not to take the stairs due to hazardous conditions caused by inclement weather (whoops), and a wooden trail marker pointing to the parking lot.

Yeah. That's where I want to go.

Immediately.

Squish.

My drenched boot squelches in time with the not-quite-so-soaked one as my frozen legs propel me forward like automatons, freshly humbled yet triumphant. I'm so close, I can see the car, and I notice for the first time how thick with snow the trees have become in the few short hours I have been gone. Their branches bend under the deceptive weight, audibly creaking and swaying in an incomprehensible, yet unmistakably musical rhythm against the roof of my car.

The juniper bows low, slender green fingers outstretched toward me, dripping with milky-blue berries, like gemstones. I realize this means I can reach them.

There are many species of juniper native to North America, in the genus *Juniperus,* family *Cupressaceae* (cypress). You might recognize the unusual, sharp flavor if you have ever partaken in a good ol' G&T, since gin is traditionally made with juniper. You may have also heard the cones referred to as "juniper berries," which is technically incorrect, but since I'm an epicure, not a botanist, both are acceptable as far as I'm concerned. All standing junipers that look like trees are edible, but creeping ornamental junipers should be avoided since they may contain unsafe levels of a toxin called *thujone.* Edgar Allen Poe is theorized to have suffered from thujone poisoning, which can cause seizures and hallucinations, in part due to his well-documented love of absinthe.

This species, *Juniperus communis,* has tacky, round, resinous cones that appear either matte blue or green and are covered in beautiful blooms of wild yeast. Its cousin *Juniperus virginiana* will develop tiny scales on each of its leaves like the toes of a lizard as it matures. I fill my pockets with handfuls of cones and clip off a few of the branches, blanketing the rest of the treasure in my basket.

Juniper is an unfamiliar ingredient in the average home kitchen. However, its culinary gifts neither begin nor end with cocktails. When fermented or boiled with a sweetener like honey or sugar, its tannic bite is quickly subdued, enriched by softer, sweeter citrus notes. When crushed or dried and powdered, it can be added to everyday marinades and spice blends for a rustic, peppery kick. The thick blooms of wild yeast that cling to juniper berries are a popular way to source interesting cultures for sourdough, beer, mead, and even sauerkraut, and dropping a few into your fermenting vessel may yield surprising results. Juniper is a fascinating tree with a wealth of stories to tell and gifts to give us, but as with most treasures worth seeking, much of its wealth is hidden away for only a curious seeker to discover. However, this wasn't always the case.

One of the secrets of this magnificent plant is unlocked in a rather peculiar way: through fire. Daniel Begay, a Diné researcher at Arizona State University, took a keen interest in the foodways of his people, and particularly in identifying the nutritional components of his community's traditional foods. As part of a project, he took samples of juniper from twenty-seven different locations on his reservation and burned them to ash (juniper ash is an additive in many Diné foods, most notably blue corn mush), then ran mass spectrometry to determine what compounds remained available. On average, Begay found that in a single gram of juniper ash, he could expect to find roughly 280 to 300 mg of calcium, nearly 25 percent of the daily recommended requirement. You'd need about a cup of dairy milk to match what 1 gram of juniper ash can do. As Western scientists learn to be more curious about Indigenous knowledge, we are beginning to understand the ingenuity and complexity of traditional Native food systems. Because a significant percentage of the world is lactose intolerant, the dietary applications of juniper ash are potentially quite significant, but to make that calcium bioavailable, the juniper *must* be burned.

***Juniperus communis*, Common Juniper**

Fire destroys, but when the smoke dies down, sometimes you'll see that the fire also makes it easier to access the things we really needed all along. What was once indigestible is now suddenly food, a barren wasteland becomes a garden, new growth thriving on the carbon and minerals that the flames left in their wake.

I sheath my clippers and slide into my car. Key in the ignition, heat cranked to high, I barely feel the cold air blasting on me as I wait for my ride to heat up. I have a pair of extra socks stuffed in my door, and some summer flip-flops in my back seat for some incomprehensible reason. After a quick below-the-knee costume change, I start to resemble my dad on a beach vacation, but at least I'm beginning to thaw out. My toes prick uncomfortably, but soon I will be able to drive.

When I return, my pipes will still be frozen, my toes will take ages to thaw, and I imagine something else will be broken when I walk in my door (with my luck, probably the door itself). But I, at least, with a bit of greenery and a few perfect mushrooms in the creel bumping gently by my side, listening to the muffled crunch of last year's leaves beneath the fresh snow and the stillness enveloping my senses, will be a bit more whole. While the new life never looks exactly the same, the energy each of us have carried through every broken-down thing is recycled and used to make something new again.

In this fleeting winter stillness, I am safe, or safe enough. I am whole, or whole enough. My life doesn't look the way I pictured it even six months ago, but the song of time never stops, and like fire, it consumes the things that no longer serve a purpose.

But if the life I lived has to burn, I can only hope to make use of the ashes.

GATHERING EXERCISE

"ECOLOGY 4:33"

Pick a place outdoors that speaks to you or seems interesting. It could be your backyard, a park, a mountaintop, or a tree in the middle of your city. Set a timer and stand or sit for 4 minutes and 33 seconds. Listen for and count the sounds you can discern. Which sounds are close, and which sound far away? How does your body feel in this space?

Wintergreen Extract

Makes 1 quart

Wintergreen has many applications, both culinary and medicinal. The Iroquois and many other Indigenous people across Turtle Island (North America) have historically pounded the berries into cakes and dried them for later use, and used the leaves and roots in teas and decoctions (simmered teas) to treat stomachaches and head colds. Today, I like to use it as a flavoring agent in cookies, cakes, ice cream, and candy. The flavor molecules don't infuse long term very well in water-based suspensions or preparations like simple syrups. However, just like vanilla beans, wintergreen's flavor lasts forever when preserved in a neutral high-proof alcohol, continuing to deepen and strengthen with time. Vodka and white rum are popular choices, but Everclear will also give you excellent results when creating alcohol suspensions. I have also found something cheaper than Everclear called Diesel Grain Neutral Spirits, which works beautifully but has a terrifying label that may bring back bad memories of poor choices made at college parties.

You can continue refreshing the jar with more alcohol and fresh wintergreen as needed. Use this throughout the year like you would peppermint extract.

An important note: *Anytime you use a jar for long-term storage, whether it be for canning, fermenting, or infusing, you should always make certain that your jar is sterilized to avoid contamination.* A dishwasher on its hottest setting is fine, but you can also boil jars, lids, and rings in a pot of water on the stove for a few minutes.

---

1 to 2 cups whole wintergreen roots, stems, leaves, and berries (any combination of plant parts is fine, but be sure to include some roots)

About 2 cups high-proof alcohol, at least 70%

Clean the wintergreen thoroughly (a vegetable brush is good for cleaning the roots). Roughly chop.

(continued)

Pack the chopped wintergreen into a quart-sized mason jar and cover with alcohol, leaving about ½ inch of headspace. Allow it to sit in a cool, dark place for at least 6 weeks, shaking at least once per week.

Do not strain—it will continue to infuse and get stronger over time. Add a small amount of the liquid (similar proportions to vanilla extract, depending on the strength) to baked goods, ice cream, candy, and more.

Miso Soup with Enoki

Makes 2 bowls

One of the most comforting things you can eat during a long, cold winter is a warm bowl of soup. Soup is a fantastic and budget-friendly way to stretch more modest rations, especially since so many are the result of cobbling together common kitchen staples and scraps (in fact, I often add leftover bits like onion skins and vegetable peelings to my broth, and you can choose to do the same here). Miso soup is one of my favorite quick and easy weeknight recipes, and you don't need to leave home to find it if you have a few key ingredients on hand. Traditionally, dark miso is used in the winter months, but I prefer a white miso because it doesn't overwhelm the delicate flavors of the enoki mushrooms.

Cleaning enoki can be a bit of a challenge since the tops tend to be sticky and hold on to dirt and debris from the environment. Try soaking them in warm water with a bit of salt for about ten minutes, then rinse off under running water a few at a time. To cook enoki (and lots of other mushrooms), I like to use a "dry sauté" method, which draws moisture out from the mushrooms by cooking them on high heat in a cast-iron pan with salt. The mushrooms will begin to stick to the pan, but rather than using oil to unstick them, you'll use a bit of water instead. I know it seems counterintuitive, but the mushrooms will continue to release water until they have reduced in size by about half. Then (and only then), you'll add a bit of oil or butter to finish them off. This cooking method results in meaty mushrooms with lovely, crispy searing and absolutely zero slime. Once you taste a dry sautéed mushroom, I promise you'll never again be tempted to drown them in butter!

- 3 to 6 inches kombu seaweed
- ½ to 1 cup vegetable scraps (such as onion skins and carrot peelings; optional)
- ½ block soft tofu, diced
- 2 tablespoons dark soy sauce
- 4 tablespoons miso paste
- Pinch of kosher salt
- ¼ pound fresh enoki mushrooms (if you can't find these, try oyster mushrooms instead)
- 1 teaspoon toasted sesame oil, plus more for serving
- 2 tablespoons chopped field garlic (you can use green onion instead if you prefer)
- Chili crisp (optional; if you like a little kick—I do!)

In a large saucepan, bring 4 cups of water to a rolling boil over high heat. Add the kombu and vegetable scraps (if using) to the boiling water and cook for 3 to 5 minutes, until fragrant, to make a broth. Remove the kombu and vegetable scraps from the broth and discard.

Add the tofu and soy sauce to the broth and cook for 1 minute (just long enough to warm the tofu through but not long enough for it to fall apart), then remove from the heat and let cool for 5 minutes.

After 5 minutes, put your miso in a ladle, dunk it in the broth, and use a chopstick to whisk the miso into a slurry. Swirl the miso in gently, taking care not to break up the tofu, until fully stirred into the broth.

Meanwhile, add a pinch of salt to a cast-iron pan over high heat and, when the pan heats up, scatter the enoki mushrooms in an even layer. Leave the mushrooms mostly undisturbed until they begin to shrink, flipping occasionally to check for sear marks. When the mushrooms start to stick to the bottom of the pan, add a few tablespoons of water at a time to unstick them.

Once the mushrooms have reduced in size by about half, add the sesame oil. Heat the mushrooms in the oil and remove before they start smoking. Split the mushrooms between two soup bowls and ladle the broth on top.

Serve with an additional drizzle of sesame oil, a sprinkling of fresh field garlic or green onion, and a spoonful of chili crisp (if using).

Juniper Ash Polenta Cake

Serves 4 to 6

Cooking with ash might seem strange, but it has a lengthy history of culinary and supplemental use, most notably in the kitchens of Diné people, who have used it for thousands of years to improve the bioavailability of certain vitamins and minerals in corn. Ash is a valuable source of essential minerals such as calcium, and because ash is alkaline, it can be used as a natural lye. Some herbs, like celery and mugwort, uptake salts that are made more available by the process of creating ash. Ingredients such as celery ash are gaining popularity in the gourmet kitchens of today, beloved for their ability to enhance and deepen flavor while imparting an earthy flair.

For this polenta cake, try using blue cornmeal instead of yellow, as the juniper ash will intensify the blue color. Polenta cake is reminiscent of a sweet, sticky cornbread, and pairs well with fresh berries and a big dollop of whipped cream.

To make juniper ash:

Collect a few juniper boughs. The best ones will be a little bit on the older side and slightly dried out (usually the needles will be tinged with red), but you can also collect fresh ones and let them dry out for a few days. Burn branches over a wood fire until blackened but still holding their shape. Let cool until safe to handle, then break into pieces and dehydrate or hang to dry until brittle. Crush in a mortar and pestle or blend in a blender, then sift through a metal sieve to filter out larger pieces until you are left with a soft grayish dust. Let cool and store in mason jars.

To make the cake:

2 tablespoons vegetable oil (for greasing)
1 cup polenta or slightly gritty cornmeal
½ cup all-purpose flour
2 tablespoons juniper ash
2 teaspoons baking powder
A hefty pinch of kosher salt
3 eggs
1 cup Grade A light maple syrup
1 cup plain yogurt (I use full-fat, dairy-free coconut yogurt)
¼ cup vegetable or canola oil
¼ teaspoon finely chopped juniper needles
2 tablespoons unsalted butter or margarine, softened
Powdered sugar, for dusting

Preheat the oven to 350°F and grease a 9-inch pie pan with vegetable oil.

Mix the polenta, flour, juniper ash, baking powder, and salt in a large bowl and set aside.

In a medium bowl, beat the eggs, maple syrup, yogurt, vegetable oil, juniper needles, and softened butter or margarine until slightly fluffy and pale.

Add the wet ingredients to the dry ingredients and mix until just evenly combined into a batter, being careful not to overmix. It should look wet and lumpy, like halfway mixed concrete.

Pour into the prepared pie pan and bake for 30 to 35 minutes, until the edges of the cake are slightly crisped and a toothpick can be inserted and removed cleanly from the center.

Remove from the oven. Turn the cake out from the pan and cool on a baking rack. When just cooled, dust with powdered sugar. Serve slightly warm. I like mine with fresh berries or a spoonful of whatever jam I have open in the fridge, plus lots of whipped cream or a big scoop of vanilla ice cream.

"Cheesy" Field Garlic Popcorn Dust

Makes 1 pint

Popcorn dust isn't *just* for popcorn—you can use it as a topping for all kinds of things, vegan and non-vegan alike! I've just found that it happens to go particularly well on popcorn, and it's a good way to process a whole lot of field garlic all at once if you find yourself the lucky recipient of a windfall (which tends to be the case with field garlic). Sprinkle some on hot buttered toast to make a quick and easy garlic bread, or try it as a topper for macaroni and cheese.

To make this recipe, you'll need to start with dried field garlic leaves. Either dig them up with a stick or a trowel, or just snip off the long, green, tube-shaped leaves with a pair of scissors. You can dry field garlic in just a few hours in a dehydrator, or if you aren't in a rush, you can tie the leaves in bundles and hang them to dry in a warm place with decent airflow (usually I do this in my kitchen). I find that the latter method takes about a week, sometimes less or more depending on the humidity in your house. You'll know that your field garlic is dry enough to use when you can easily crush the leaves into a powder with your fingers.

2 cups lightly packed dried field garlic leaves
½ cup nutritional yeast
1 tablespoon fine salt
A desiccant packet (optional)

Working in batches of about ¼ cup at a time, place the dried field garlic leaves in a spice grinder and grind until you are left with a fine powder. Add each batch to a medium bowl until all the field garlic leaves have been powdered and added to the bowl. Repeat this process with the nutritional yeast, grinding to a powder in small batches and adding to the bowl of field garlic powder. This will create a uniform end texture for your spice blend.

Add the salt to the bowl. Using a fork or a whisk, toss together until all the ingredients are thoroughly combined. Your end product should be an evenly distributed greenish-yellow color.

Store in a jar with a desiccant packet to minimize clumping. If you don't have a desiccant packet, shake well to reincorporate before using.

Sprinkle generously over well-buttered popcorn, or a dish of your choosing.

The Creeping Green

The snow is melting into music . . .
the deepening murmur of the awakening water.

John Muir,
The Mountains of California

APRIL 2015

Indianapolis, Indiana

Where the hell are you, you little bastards?!

I'm somehow cold *and* sweaty at the same time. My filthy body aches from crawling around on the forest floor, searching unsuccessfully for an elusive treasure. After nearly two weeks, I'm about ready to give up, but I'm feeling the weight of the sunk cost of this ridiculous endeavor.

Why was I wasting my entire weekend scraping through brambles and pulling twigs out of my hair? For something that must, by all accounts, be genuinely wonderful, though I feel as though I've been sent on a snipe hunt.

It will be getting dark soon, and I should be getting home. Another empty basket day. I stand up, knees creaking a bit more than usual, feeling rather stupider than my ego would typically permit. A sea of trees whose names I do not know ripples beyond my feet, and I press through their arching bodies to return home.

MORELS ARE A holy grail of the mushroom foraging world. Even inexperienced foragers who don't know their boletes from their polypores still glow a bit at the sound of the name, and willingly sacrifice countless hours in search of these generally diminutive, decidedly odd fungi. Are you *really* even a mushroom hunter if you haven't found your first morel peeking through last autumn's leaf litter, smelling of rich, freshly moistened soil and an indescribable umami perfume?

Well, I'm not about to miss out again. I've spent two spring seasons scanning the earth for morels, to no avail. Either I don't have the knack, or I'm doing something wrong. My eyes sweep the ground before me, left to right and back again, looking for any hint of inconsistency along the path. Nothing but last year's decomposing maple leaves, green plants pushing impatiently through the cracks between rocks and rotting logs, and a few empty beechnut casings.

A mushroom catches my eye—nothing like a morel, but a mushroom nonetheless, flat as a platter, with gently rounded margins and a vaguely scabby-looking brown and tan cap. I get closer, taking a brief hiatus from my obsessive searching. This looks like a pheasant back mushroom, also sometimes called a dryad's saddle, known by the scientific name *Cerioporus squamosus*. The word *cerioporus* comes from the Greek *kerion*, meaning "honeycomb," and *squamosus* is a Latin word meaning "scaly." Sure enough, a quick glance under the hood reveals a creamy white pore surface, the aggregate of which looks like a honeycomb on closer inspection, and the feathery-ticked cap does indeed appear scaly. Edible, technically, or so I've heard, but not *choice*. Not a *morel*. I sigh, crack my neck, and continue on.

Finding your first morel is *hard*. Most people give up after the first year of hunting, and part of the reason is that most of the information about how to find morels online contradicts itself. According to Google, morels prefer both damp, loamy soil and sand. Also, pine duff. They grow in association with dying elm, but may also be associated with the Dodge Caravan sitting in your driveway. They might be mycorrhizal or saprotrophic, or endophytic, or possibly all three. You should always cut them to avoid disturbing the mycelium, but research shows that plucking them makes no difference. False morels are lying in wait to trick you, but apparently shouldn't be called false morels because they look nothing like morels, unless you're inexperienced, in which case you cannot trust the evidence of your own eyes and shouldn't be out looking for mushrooms anyway, because you'll probably accidentally kill yourself.

Frankly, it's easy to get overwhelmed. Where do you even start with that information? And more importantly, how much of it is actually valid?

The answer, as with most things having to do with mycology, is a bit overcomplicated.

Morels belong to the phyla *Ascomycota*, or sac fungi, in the order *Pezizales*, which contains a wide range of appetizing specimens including *Peziza domiciliana*, which you might know better as "damp frat house carpet shrooms." Also in the order are the scarlet elf cups

Sarcoscypha spp., and even the so-called false morels *Helvella, Verpa,* and *Gyromitra* spp. Morels themselves (*Morchella* spp.) comprise over eighty described species at the time of this writing, and likely by the time you read this will be well over the one hundred mark. Their life cycles and patterns are complex, unpredictable, and subject to minute changes in factors like ground temperature, disturbance, associated organisms, moisture, soil composition, and more.

On top of that, their grayish-brown, deeply pitted fruiting bodies blend in with last year's leaf litter. Such an exasperating habit. No wonder they're tricky to find.

With fungi, we're stuck with a constant "chicken or the egg" problem, but for the sake of discussion, let's start the life cycle of a morel with the egg (in this case, microscopic spores released from the deep pits of mature morels). Spores are commonly dispersed by wind, rain, and animal activity, and if all goes well, will settle on a preferable substrate within a range of suitable trees for its species.

To understand the growing habits of fungi, we need to recognize that not every fungus receives nutrients in the same way, fits the same ecological niche, or prefers the same type of habitat. Some mushrooms, for example, are what we call *mycorrhizal,* meaning they grow in mutualistic association with living plants, typically tree roots. The iconic red and white spotted *Amanita muscaria* is one such mushroom, bargaining nutrients such as nitrogen and phosphorus in hopes of an exchange for its favorite snack, carbohydrates. (I can relate.) These fungi develop intricate relationships with their tree partners, often building complex, expansive underground carpets of silken threads called *hyphae,* which make up mycelium, an organism that resembles the tree root systems they are entangled with. If a mushroom is like a piece of fruit hanging from a tree, then the hyphae is a single branch, and mycelium is the tree.

Other mushrooms do not behave this way; instead of acting as merchants, bartering with living trees, the fungi we call *saprobic* are more like vultures, the death eaters. They are decomposers, breaking down dead plant matter and scavenging for morsels of energy. Without saprobes, we would have no flow of nutrients—carbon and nitrogen couldn't be cycled through the soil or decay. Organic material builds up quickly, like trash bags on the sidewalk, and if nobody comes to pick them up, eventually you won't be able to travel anywhere. Every fungus has an ecological niche, but like a trash collector, they often go unnoticed until their work isn't being done.

Most morels exhibit both mycorrhizal and saprobic habits, as the mycelium is signaled to fruit when in the presence of decaying, nutrient-rich organic matter provided by dying trees, but they begin their growth process by knitting their hyphal threads underground in a web through a favorable area full of living trees. Some morels even develop sclerotia, dense masses of knotted mycelium formed beneath the soil as a precaution, to store energy through challenging environmental conditions. Others are carbon-loving "burn morels," also known as *phoenicoids*, known for their ability to rise from the ashes like the legendary bird and remediate soil after wildfires tear through the landscape. A clever morel hunter on the West Coast may strike gold if they simply seek out burn scars when looking for morels, often finding themselves among an abundance of mycological treasure far beyond any one person's ability to harvest.

Morels are an example of the rewards of patience in the natural world, and not just because of the immense effort most people go through to find them. If conditions are unfavorable, the mycelium will not expend energy making fruiting bodies. Similar to a very particular artist, the mycelium's work will not leave the shop until it can be well-received by the environment it lives in. Understanding the morel's tendencies and preferences is key to learning about where to look, and how to maximize your chances of locating them in the right place at the right time.

Breathe in. Hold. Breathe out.

Humans are often oblivious to many of the intricacies of the world we live in. I remember being ten or eleven in my home forest, walking past and sometimes crushing plants without knowing them, half the time not even *seeing* them. Ecologists and botanists refer to this as "plant blindness," a recently described societal condition characterized by the inability to perceive individual plants in one's own environment; in other words: an immunity to *noticing*. Morel hunting will put all your observational shortcomings on full display, as it has for me time and time again.

To be fair, it isn't just plants or frustrating little fungi that I've missed. Lately, I feel as though I don't have time to notice *anything*. Sometimes I get to work and I have no idea how I got there, or I realize it's nearly half past two in the afternoon and I'm absolutely incapable of recounting any of the day's events. Weeks, months, and entire seasons pass before I've even had the chance to acknowledge the time. I'm trapped in a routine of all the next "I-have-to's" before I can even begin to think about the immediacy of life. A world of sensory experience is lost to me, blurred around the edges as though I've left my glasses on the nightstand.

It dawns on me that I am halfway through my twenties with very little to show for it. Who was I before the world trapped me in its heavy spin cycle of school loans, electric bills, side hustles, frozen dinners, and tax forms? These days, I can't remember most of the things I've done in any real detail. *Are most of my experiences really that forgettable?* I wonder, horrified at the prospect. *Am I wasting my life?*

I have to do something different, I think to myself, trying to find a way to be present in this moment without urging the next one to begin. I remember how I would try to make myself fall asleep when I was small, on the advice of a fellow insomniac, fully relaxing first my feet, then my calves, and each subsequent part of my body upward and upward until I reached my neck. The exercise more than occasionally worked, which I attribute in no small measure to the uniquely childlike mysticism that causes us to believe that our ordinary actions have the potential to achieve extraordinary results. Perhaps, though, it worked not because of my sincere belief that it would, but because it reconnected my ever-preoccupied mind to my corporeal self. Instead of *watching* myself like a somewhat bored television viewer, I was actually *being* myself, repairing a connection that should have always been there. I try to focus on my steps, the sound of leaves underneath my shoes, my calves tensing as I climb uphill.

Foraging puts me right in more ways than one, but perhaps one of the most obvious effects of time spent meticulously observing the woods is how easy it becomes for me to identify who I really am and how I really feel. The singularity of the hunt focuses my mind, magnifying my awareness of my own body. Alone in the woods, with twigs in my hair and mud on my knees, I'm not concerned about what I look like. I'm just . . . *part* of it.

When I arrive home, I open my laptop and return to the forums, preparing myself for more contradictory hunting advice from hopefully well-meaning strangers when I stumble upon an infographic showing five different species of trees associated with morel species. *This could be the missing piece,* I muse, jotting a few notes down in my phone.

Until now, I had only been searching for elm trees, but they had been quite difficult to locate due to the devastating effects of Dutch elm disease, *Ophiostoma* spp., a fungal pathogen that causes the tree to plug up its own water-uptake systems, eventually starving it to death. The stowaway disease was transported to North America via beetles hiding in European elm logs in the 1920s, where unsuspecting craftsmen would accidentally unleash it upon the vulnerable American elm, reducing its numbers by over 40 million and counting. At one time, nearly 75 percent of all the elms in North America were affected. For many years, large flushes of morels bloomed in the wake of massive tree death, but as the trees continued to disappear from the forest, the nutrients that fed the morel mycelium also began drying up, causing the peculiar little fungus in my area to become somewhat scarcer.

Scarce, yes, but not impossible to find. Perhaps it was time to expand my search and consider consulting with other beneficial host trees. I close my eyes and as I fall asleep, I see spinning tessellations of rotting leaves and things that look like morels but are, infuriatingly, not.

MORNING AGAIN. I force my eyes to open as the rose-pink sun crests the tree line outside of my little house on the northeast side of Indianapolis, and realize I left the window open. The air is warm—it must have caught the fists of last night's chill, leaving evidence of the tussle in the form of glassy dewdrops and a haunting mist settling on the lawn. I absently dress myself, run a toothbrush through my mouth, tie my hair back, wrench my sneakers on, and burst out the door. Nothing matters but the hunt.

My first stop is a grove of tulip poplar, *Liriodendron tulipifera*, an absolutely gorgeous member of the magnolia family adorned with pendulous orange flowers from April through about June. These trees are mostly bare except for the beginnings of green leaves, but according to my research, tulip poplar are a preferred partner for the petite Midwestern morel species *Morchella diminutiva*. Assuming the "mushroom hunter's stance" (an arthritic posture in which one's body is hunched as low as possible in order to make it easier to see terrestrial mushrooms), I wade through the vegetation, searching for any sign of the small, wrinkled mushroom that has become the subject of my obsession.

Mushrooms, almost all of them, like to hide. For every one you see, there are a dozen more just outside your vantage point, hiding beneath logs and leaves, shielded from view on the other sides of trees and rocks. Without adjusting your perspective, you'll miss them every time. Look up high, get on the ground, lift the brush, walk circles around the trees like they owe you money. Mushrooms respect thorough examination, and they'll reward you accordingly.

The footpath is well-worn but decaying as nature reclaims it, little rivulets of snowmelt from the rocky cliff to my right carving deep fissures in the earth. For some reason, my eye pauses on the dark green plants dotting either side of the path, broad leaves low and almost greasy with ragged edges and veins like lightning scars. I recognize them—garlic mustard, *Alliaria petiolata*. A noxious European invasive in the mustard family with pungent, spicy-garlic leaves (hence the common name), a horseradish-flavored taproot with a bright purple stain in the center when you snap it off, it has an unfortunate habit of spreading rapidly through forested areas, releasing allelopathic chemicals in the soil and damaging native plants' chances of survival. Garlic mustard may not be welcome in this forest, but it is certainly welcome in my kitchen, where it will be pounded into pesto, blended into salad dressing, and fermented into kimchi. I start coaxing the plants out of the ground, knocking the roots to dislodge the rich soil so something less destructive can hopefully move in. *Come on, I'll take you home with me.*

The baby plants surrender willingly for now, but I know that within a few weeks they will grow tall, sprouting white flowers and burrowing their roots deeper into the ground in their adolescence, making them harder to evict. I cannot possibly pull all the garlic mustard in front of me before moving on, but I make a decent dent in it, lining the bottom of my basket with green leaves and fat taproots.

***Allaria petiolata*, Garlic Mustard**

Before me is an occasional floodplain, which on this occasion happens to be mostly muddy clay. Black cherry and ash trees huddle together, spindly branches forming a hollow canopy. This could be good hunting, even though I'm bound to get filthy. A small price to pay for the hope of treasure, though.

I take care not to trample the mottled leaves of trout lily, *Erythronium americanum*, delighting in their absurd, tie-dyed complexion. While edible, trout lily leaves should be eaten sparingly, lest they exit the same way they entered (along with the rest of your lunch). I decide to risk it, plucking a large one and letting the refreshing cucumber-sweetness flow across my tongue. Amid the trout lily, clumps of sochan, *Rudbeckia laciniata*, are beginning to emerge, dark green leaves unfurling, each with a thick, juicy, vegetal stalk in the center. I snap a couple of the fattest stems and begin to munch, absentmindedly enjoying their fresh celery-parsley-cilantro flavor.

Weaving in and out of trees, examining the ground for any sign of morels, I am left stumped. They should be easy to spot here, with only a thin layer of leaf litter resting on almost perfectly flat ground (perks of living in Indiana). Still, nothing. The trees wave down at me, clicking and tapping their branches together in greeting. Though I do not know what they call themselves or one another, I can recognize them.

I've always had a hard time understanding what other people are thinking, how they're feeling. Trees are easier. They tell you who they are, they wear it on their skin, in the green of their hair, in the flowers and fruits they adorn themselves with. People are mysterious: unless they're wearing tags to tell me who and what they are and how they feel, I'm in the dark. A tree is the sum of all the other trees it has ever been. With each new season and each new ring, the echo of the past grows fainter.

I wish I could be a tree and grow new skin over my old self every year until I could forget this terrible year. Loneliness feels like it lives in my DNA. I rarely see my spouse, or perhaps more truthfully, rarely feel seen *by* him. Like socks that have stretched out from overwear, I can feel the fibers fraying and holes beginning to form in our communication, the heel spinning uncomfortably to the top of my foot with every step I take. When I try to share my curiosities and ideas, the benefit of the doubt never seems directed toward me. I've learned that it's easier not to talk. Out here, I don't have to explain. I'm allowed to be sure of myself, to fade outside of one reality and into another with every deepening step into the woods.

My stomach growls, reminding me that I missed my only opportunity to eat breakfast in my haste to venture out. It's too early in the season to find any substantial snacks to munch on in the forest—no berries, no wild plums or crabapples. Oh well. Time to keep moving. My stomach can wait. I'll fill it with morels, if I can manage to find any.

My bones still seem to be shaking off the chill of winter, missing the buoyancy that will eventually come with warmer temperatures, when the forest becomes a banquet of lush green and the fragrance of new flowers. I have yet to truly awaken from the darkness-filled winter stretched out behind me, a shadow that looms large in my periphery. I am, for the moment, frustratingly liminal.

I think I need to take my shoes off. Something nobody will tell you about hunting for morels is the extreme subtlety of their preferences, even down to the looseness of the soil and how far your toes can dig into it. Morels can grow anywhere, but they like soil with some generosity, a bit of sink. I pull my shoes off, tying the laces together in a bow and slinging them around my neck. I do this often enough that sometimes I wonder if there is really any point to putting the shoes on in the first place. The ground is cold, but my body is finally waking up from the associated contact. Gravity coaxes my toes down into the damp leaf compost and rich humus, and I sigh at the pleasant, nostalgic feeling.

"Wipe your feet before you come back inside!" I hear my mother shout as my brothers and I race outside, raindrops and muddy feet pounding in time on the gravel driveway. The sweet perfume of petrichor and mountain laurel are lush and overpowering, and graceful white flower buds are beginning to emerge from the tips of the blueberry bushes. My brothers, Christian and Zachary, chase each other like squirrels through the backyard, shrieking with laughter (which will likely turn to tears when Christian inevitably starts playing too hard).

The mossy logs and jagged slabs of lichenous slate poking out of the ground are slick with rain, but occasionally, a flash of bright orange will appear upon one of them. Red efts, amphibious juvenile newts around 1 to 3 inches in length with neon skin, are flooded from their holes in the ground in the spring during weather like this, and my brothers and I never tire of finding them.

Normally an easily distracted child, I can nevertheless spend hours looking for them, meticulously checking every rock, tree trunk, and fallen log.

I am lost in thought, wandering aimlessly for so long that I have forgotten that I came here to look for mushrooms. With my left foot, I brush something that makes my heart stop. *A snake*—no, wait. When I was a teenager, a snake slithered across my bare foot when I wasn't paying attention to where I was going—they are shockingly cool and smooth, with a momentarily frightful weight to them. This was just a snake shed, perhaps ten or eleven inches long. I pick it up and examine it. It's transparent, hollow, delicate as a leaf skeleton, artfully formed. It must have survived the winter, as have I. No telling whether the snake made it, only that it discarded what no longer suited it.

I've been noticing that my memory isn't so good anymore. Maybe I *am* like a tree, and the older rings are buried too deep under my new skin. Too much time, too much distance, maybe just too much disappointment between here and there. I avoid thinking about it, if possible, but I find myself wanting to escape back into those happier memories, wishing my life was still as uncomplicated as it was back when my main concern was finishing my chores so that I could be released from the house to look for newts.

Maybe I don't want to be a tree, growing new skin over the old until the past has been buried under layers of wood. Even with the passage of time, the past is still with me, within me. Being a snake seems more expedient. Maybe I don't want time to *bury* my regrets, but rather to shed them completely. Start over fresh, leave my old life behind while I go look for something better. *No,* I sigh. I'm being unfair. Humans don't get to shed our skin; we have to live with the consequences of our choices. I'd probably find a way to make a lot of the same mistakes over again, even with the benefit of hindsight. Sometimes you can't know what something will become until you've already taken a chance on it. The surface of the water can hide a multitude of dangers.

LOOKS CAN BE deceiving, both in life and in foraging. Over the years, I've spent a lot of time and effort gathering food that turns out to be less than what I hoped it would be. Sometimes the biggest, fattest grapes are hopelessly sour, occasionally a giant puffball turns out to be yellow-green in the middle despite being perfectly white and firm on the outside. Even experience won't always save you from spending hours performing what may very well be fruitless labor. Maybe it doesn't get any better. Yet somehow, I'm still out here. It still feels worth the effort to try, to hope that I'll eventually find something that will delight, not disappoint.

The snake shed his too-small skin and found himself better for it.

I still wish I could shed my skin.

I bury the uncomfortable thoughts once again, and after a moment's hesitation, pick up the snakeskin and tuck it into my basket, sprinkling a few flower seeds from last season in exchange. Perhaps the snakeskin will bring me good luck.

I am descending slightly, and the ground is wetter now. I feel the cold sludge welling up between my toes, and I catch a fleeting scent before the wind changes. It smells bright green and skunky, like garlic and new grass. It is familiar, exciting. I switch gears, scanning the floodplain eagerly for any sign of the plant I know belongs with that scent, and then it hits me all at once: I've been walking through a patch of them for at least the past ten minutes. The bunny ear leaves of ramps appear before my waking eyes in every direction, clusters spreading for acres, as far as I can make them out and farther. Morels forgotten for the moment, I pounce on this unexpected gift. I select the broadest leaves, collecting one at a time from the biggest plants, tucking in the garlic mustard at the bottom of my basket with a fresh blanket of glossy green.

Ramps, or *Allium tricoccum*, are big business in the culinary world, and a source of bitter annual controversy among passionate foragers. The plants grow slowly, taking around seven years to fully mature, and removing the whole plant will prevent it from reproducing. The question punted around in foraging-centric spaces each year always seems to come down to "can they be harvested sustainably?" Should you dig, cut, and leave the root base, take one leaf per plant with two or more leaves, harvest 10 percent of the patch, or leave them alone altogether? Opinions vary widely, with historic cultural traditions seemingly at odds with the available science, which up until recently has painted a damning portrait of bulb harvest.

Disagreements about what constitutes an ethical harvest have ended friendships. It doesn't help that commercial foragers along the East Coast can regularly be spotted in early spring with shovels, buckets, and trash bags, making quick work of entire patches to sell to restaurants and at farmers markets and specialty shops. Chefs salivate over the short-lived springtime delicacy, enchanted by their complex, crisp greens and pungent bulbs, buying them at a premium. It's somewhat unfortunate that this ubiquitous wild food has become so deeply entangled with commerce. On the one hand, more people get to learn about and enjoy them, but on the other, dollar signs can be blinders, incentivizing the forager to behave with reduced temperance.

Many foraging educators recommend collecting only leaves from a small percentage of an established patch, ensuring that each plant that experiences any harvest is left with at least one leaf to allow it to continue photosynthesizing. This makes good sense, particularly when applied to smaller, more sparse populations of ramps, and it has the added bonus of teaching newcomers to be thoughtful and restrained across the board when interacting with wild food. In order to avoid overharvest, some foragers impose numeric limits on their collections, such as "no more than 10 percent of the patch." On private land, this may be sufficient, but on public land, with enough well-meaning foragers following the same advice, a small patch may easily be decimated. That's just math.

On the other hand, this advice might actually be less useful when applied to the dense carpets of ramps spanning multiple acres that can be found throughout the upper Midwest. According to research performed by respected author and wild food expert Samuel Thayer (a man who ranks among my most dynamic, absurd, and cherished foraging friends, and who may on special occasions be witnessed wearing a banana peel as a wig), it is possible that everything foragers have believed about ramp harvesting for the past twenty-odd years is wrong, or at least overblown.

Sam's research reveals that these upper Midwestern ramp populations are bigger and more robust in recent years than they were a hundred years ago. He observes that ramp plants are particularly good at "filling in the gaps" in the forest floor, and that their tendency to cluster and crowd is evidence that appropriate management of these plants *should* include thoughtful bulb thinning in favor of encouraging more biodiversity on the forest floor.

There is palpable tension between these perspectives. On the one hand, the implication that human beings should never touch or interact with the rest of nature for fear of our destructive tendencies is deeply problematic and has been used violently against subsistence cultures for hundreds of years. It is not only infantilizing, it is anti-scientific. On the other hand, there is some evidence that commercial pickers have made a significant dent in once-flourishing ramp populations up and down the East Coast, egged on by a "ramping up" of demand by consumers desperate for a taste while they're in season. From both perspectives, human greed must be accounted for.

Nuance. Thoughtfulness. Two tenets that are difficult to communicate and even more difficult to live. Approaching topics like ramp digging with nuance and thoughtfulness is key to centering what is important, to actually *observing*. Given the information we have, it would appear that the only real way forward is to honestly and fairly read the landscape before us, to identify what we can reasonably take without doing harm, and in cases where there is abundance, to collect in the best possible way to encourage the health of the patch, focusing any bulb-pulling within the densest portions. Not every patch is identical, not every biome has the same needs or composition. Forests are not ruled by simple math that works perfectly in every situation, because not every forest is the same. They are dynamic systems that are ever-fluctuating and changing, demanding something far more expensive than a fixed calculation of restraint: they need our *attention*. What does the *plant* need in this specific context, what does the *forest* need in this situation, and finally, what do *I* need for my table? All needs must be honored, all needs held in balance.

A FEW YEARS ago, I spent several weeks in Guna Yala in the Darién Gap, on the unceded land of the Guna people within Panama, through a residency program with several other artists from around the world. Each of us had different artistic practices—some were painters, others sculptors, dancers, or fiber artists, and each of us had different ideas about what we wanted to learn from the experience, but my goals were to collect audio recordings of local wildlife, identify local flora and fungi, and to learn from the locals about the traditional foods and medicine of the region. We were given unprecedented access to so many beautiful aspects of the Guna culture, from ceremonies to crafting workshops and lessons on mythology, but the local plants and mushrooms were unfamiliar to me and I craved more information about them. The jungle was a dangerous place to wander alone, and without a guide to help me, I couldn't properly speciate much of what I saw.

After some weeks, we were finally permitted to follow one of the local elders, known as *silahs*, as he ventured into the jungle to collect medicine for the community. This silah was a locally renowned herbalist named Manuel, around sixty years old, standing no more than five feet tall and armed with a tattered burlap sack and a machete. Following him proved to be more challenging than any of us could have expected, as he zipped along at top speed despite his age and size with the rest of us huffing and puffing behind him in our galoshes.

Manuel sliced his way expertly through the overgrown jungle, revealing a path that only he could see. He rarely stopped to explain himself, relying instead on a string of translators to convert his actions from Guna to Spanish, and then finally to English in a comical game of telephone. Half the time, the apologetic translators had no idea what the plants were, only "this is good for pregnancy," "the goo from this plant helps with eyes," "we use this for stomach pains." Plants were approached, gathered, and on he went, humming to himself. He hacked at bark and branches without hesitation and placed them in his bag mid-stride.

"Do you ask permission before taking the plants?" I worked up the courage to ask Manuel, trying to reconcile what I was observing with what I had learned about respectful harvesting from the Indigenous people back home in the United States. He seemed confused, and I explained that I had been taught to always ask permission from the plant before harvesting anything. He looked at me as one might look at a rather foolish child, stifling a grin. "Ask? I don't ask, no. I know them. I sing to them and let them know how they will be used. We don't have to ask, they are happy to give themselves to us. It is an honor for the plant to be used as medicine for the people, and we honor all the plants through our mola [intricate embroidery] and our ceremonies. They are grateful to us, and we will always care for them."

This behavior that I had observed was not, as it had appeared at first, a sign of disregard, but a matter of difference in how respect was afforded and how ownership was understood. I realized that what I thought had been Manuel humming to himself was, in fact, his singing to the plants as he gathered them. It was a stewardship borne from his deep friendship with the plants, a friendship built on mutualistic trust and long-standing familiarity. Manuel knew the plants well enough to know how they would want to be treated without asking, just as you probably know how your partner likes their coffee or how they like the dishwasher loaded. In Manuel's culture, family is always welcome to whatever you have. Ownership is a value of commonality, not pure individuality. The plants understand that they belong to the mushrooms and the animals just as much as they belong to themselves.

On our way back, we stopped in one of the fincas, areas designated for farming fruits and vegetables. It was hard to tell where the finca began and the jungle ended, as the lush chaos simply continued on. Plantain and pineapples, mango and achiote, lime trees and coconuts, flowering plants I newly recognized like achiote and spider lilies, all planted in a haphazard tangle with twinkling blue-eyed day flowers and pale-husked ground cherries. "We plant the things we eat so that we will always have them, but we plant enough for the animals too. The animals come and eat the food we provide for them, so we do not have to travel as far into the jungle to hunt." Manuel retrieved green coconuts and hacked them open with his machete, offering each of us a long drink. I sat at the base of the mango tree, marveling in silence at the sheer number of birds, insects, and small mammals that came to feast. Coming from a world of deer fences and bird netting, it was strange to see a garden that welcomed rather than discouraged animal visitors.

WHEN YOU OBSERVE nature, you can choose to see rugged individualism, or you can choose to see mutual flourishing.

Settlers who arrived on ships from Europe were not primed to recognize or understand this balance. The places they came from were not built for mutual flourishing; they were built for metal and agriculture and the relentless pursuit of land and money. In this "new world," they saw an opportunity; not to connect with the land or its people, but to conquer it and claim its resources for themselves. Pelts for clothes, wood and stone for houses, gold and silver for lining pockets with endless riches. The settlers shaped the new world in their own image just as they had done to the old, like hungry ghosts who consumed with no hope of satisfaction, and taught their children to do the same for centuries.

This mode of consumption and carelessness led to the ruin of much of America as we know it, but old habits die hard. Even attempts to reclaim nature were stained with white supremacy, claiming that *humanity* itself must be at odds with the natural world, and thus, nature must be protected from the spoilage that inevitably comes from human interaction with it. The rotten truth is, in fact, that the core of the conservation movement in the United States is eugenics and white supremacy, and these founding conservationist ideals rejected Indigenous stewardship in favor of a more hands-off approach, except of course where it favored the desires of those in power. This perspective was predicated on the idea that the natural was meant to be *appreciated* with the eyes, but not *embodied* in the everyday, and certainly not maintained through regular interaction. That nature was out *there* for us to occasionally look at and take resources from as our right, not in *here*, the fundamental basis of who we *are*, no matter how much of it we mow down to build our metal cities.

It's easy to see why wealthy, white landowners in the nineteenth century assumed that the natural world was a deistic machine that would surely thrive best when allowed to manage itself, unhindered by what John Muir, in his book *My First Summer in the Sierra*, called those "dirty, backwards Natives." What they could not understand was that, not unlike their rectangular rows of vegetables and herds of cattle, the wilds of America were far from untouched, and far from untended. These newcomers could not comprehend the vast community gardens cared for over many generations, towering hickory and chestnut orchards, lakes brimming with manoomin (wild rice), and acres of blueberries maintained with gentle fire, not fences. Even if they could understand all that, they would be bewildered by the notion that any garden should feed wild animal relatives as well. These lessons can only be learned when one has shed the idea that might is right, every man is an island, and food is scarce.

On the other hand, the *result* of the forcible separation of humanity from our animal, plant, and fungal brethren has bred centuries of devastating ignorance, leaving many of us effectively bleached out by colonization, lost children with no ancestral knowledge of our land and how to care for it. When we are disconnected from the vast network of living things that surround and inhabit us, when we do not possess community with one another, all we know how to do is consume. We cannot, as my Indigenous siblings here in North America would say, "walk in a good way," if we have only learned to hoard as much as we can of what is valuable to us. We cannot be good relatives if we believe on some level that there is not enough to go around. Our greed will compel us to take more, and more, and more (just in case), and we will never be content. Are most of us too clumsy and foolish to know how to behave any differently?

Every forager will, at many points in their journey, encounter the puzzle of determining how much is too much to take, or perhaps, more importantly, if there is a *better way* to take? It seems that we all agree there should be limits to our harvests. Should we reduce it to a percentage, say, never taking more than 10, or 15, or 25 percent of what we find? Should we avoid harvesting the roots of plants in an attempt to reduce harm? Where does a symbiotic and give-and-take relationship become parasitic?

Dr. Robin Wall Kimmerer's approach to the Honorable Harvest, while offered kindly, is a firm reminder that as humans who walk upon this earth, we must consider the broader implications of our actions, not only for their direct impacts, but also for what our actions signify. The Honorable Harvest is a simple but powerful set of guidelines, focused on bringing us into a balanced relationship with the land and with one another. Its foundational goal is relational reciprocity; that when a gift is received, the cycle of giving must be continued in kind. Hands that receive must also practice giving. The land that sustains us must *be* sustained, or else it will run dry and fallow.

This mindset is not *only* applicable to foraging. Rather, it is a realignment of how we consume and the respect we give to the land and to one another. It is a call to reciprocity, restraint, and gratitude in *all* things. Taking, whether a gift of fresh bread from a neighbor or a library book, should always be a reminder of our responsibility to return life to the places from which we take life to sustain ourselves—a reminder that nothing is truly free, even when freely given. A gift is an invitation to relationship, not exploitation.

We are creatures concerned with making meaning, and even the smallest actions reinforce how we value and respect the beings around us. Breaking the cycles of overconsumption by using up what we have and not taking more than we need makes our footsteps lighter. Wasting the gifts we have been given by chasing trends, being careless consumers, and exploiting the poor for our own gain will quickly throw the world out of balance. Foragers in the process of developing a harmonious, reciprocal relationship with their environment (hopefully each and every one of you who reads this book) should adopt the mindset of the Honorable Harvest.

Guidelines for our interactions, like developing interspecies trust through asking, exercising thoughtful restraint, and practicing gratitude, are not only good for the environment that surrounds us. These relational boundaries are also good for *us*. Entering into reciprocity with the flora, fauna, and fungi that sustain our lives invites kinship, kindness, and purpose, and makes us better stewards of the land that holds us. Living with the tenets of the Honorable Harvest in mind is a safeguard against selfishness and fear, offers us a pathway to the benefits of mutualism and community care, and encourages us to rise above the capitalistic ideals we are taught to idolize through consumerism. The economy of gifts is one of relational interdependence, a break from the shortsighted goals of the here and now. It is through this model of relationship-building that the balance of good life is not only established, but maintained.

Perhaps, if the "fathers of conservation" had been open to the soft ministrations of the Honorable Harvest knocking upon their hearts, they would have been able to deal more respectfully with all their relations too.

THE RAMPS HAVE stained the thumb and index finger on my right hand with their cloying perfume and green ink. I reek, yet *deliciously*. The petrichor, peat awakening from torpor, tacky pine resin clinging to my feet, and snappy fresh ramps have all coalesced on my skin, washing whatever had remained of my human scent away. I experiment with sitting perfectly still beside my basket and letting the forest swallow me.

I don't wear shoes. I never wear shoes. I have leather feet, and I run on gravel. It can't hurt me. "I am in this forest and this forest is in me," I am chanting to myself in the warm glow of my own imagination, lying in a bed of moss six inches deep beneath the oaks as rain drills into the thirsty floodplain behind my house.

I can feel myself being watched by tiny eyes as I lie alone, motionless, but I am not afraid. The animals in this forest seem more comfortable with my presence than before. Perhaps I have simply applied an acceptable level of sylvan cologne to mask my human funk.

The newts are emerging from their mud huts, flooded out for now until the trees have drunk their fill. They are not afraid of me, nor do they see the rain as an inconvenience or catastrophe.

A whitetail doe—delicate and small, perhaps born last spring?—wades confidently through the sea of ramps, snuffling the ground about ten or fifteen feet from me. Deer won't eat ramps, but there are plenty of other creeping greens for a hungry ruminant to enjoy. I suck my teeth and avert my eyes apologetically, wanting to avoid appearing like a predator, but she doesn't seem to notice me, or if she does, she has already judged and found me harmless. She nibbles at the scrubby underbrush, unhurried, while I unfocus my gaze, my body and mind slipping into a state of blank neutrality. The rhythmic clicking of branches above my head catches my ear, as black cherry trees like gothic chapels tease the soft grays of oak and maple.

I am in this forest.

(Breathe in. Hold.)

This forest is in me.

(Breathe out.)

I don't know how long I've been sitting here. The deer is so close now that we can hear each other breathing. I won't look at her eyes for fear of spooking her, so I watch her cloven hooves instead.

Something no one will tell you about morels is that they tend to appear only after you stop looking. If you move too fast, even with a trained eye, you could walk past hundreds of them without having a clue what you're missing.

Resting is the key.

She takes a step, and then another. Closer. *Closer.*

There is something to be said for an entity that demands patience of us before it will reward us with its presence. I catch my heart in my throat, blood rushing my ears with anticipation. To the left of the doe's hoof, a shape I have only seen in two dimensions, yet so obvious I'm baffled that I could have missed it.

Morels are dainty, secretive. The first morel can be difficult to spot at first due to the sheer number of decoys, but once you see one, more are revealed as if by some ancient magic.

Waiting for the doe to move along is suddenly an excruciating process. My eyes are fixed on the morels, and I want nothing more than to reach out and pluck the funny little gems from the earth, but the doe persists, nibbling on tender trillium greens.

She looks at me suddenly and I shift my gaze, meeting her enormous eyes without thinking. The patchy hair on the ridge of her spine stands on end.

The moment broken, she snorts once and leaps away in a flash and a flurry.

You won't always leave the forest with a basket full of food.

More often than not you'll come up empty, nothing to show for your time and effort but scratched-up extremities from crawling through brambles, assorted insect bites, leaves in your hair, and a basket full of inconsiderate people's busted beer bottles and pop cans. But I have to think that time spent among trees and soil is always time well-spent, dinner or no dinner. The leaner trips make you appreciate when nature is so generous to your table that your triceps ache from carrying your basket, the kind of abundance where you could not possibly eat alone and risk hoarding all that joy.

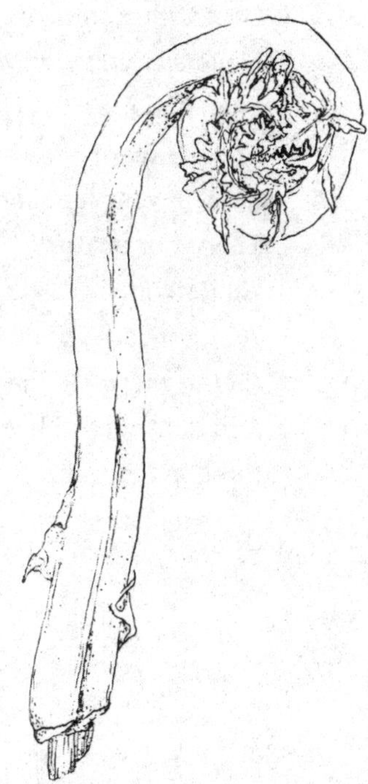

***Matteuccia struthiopteris*, Ostrich Fern**

I am floating, unable to stop admiring the

perfect morels laid to rest on a blanket of ramp leaves, trout lily, garlic mustard, and a sprinkling of ostrich fern fiddleheads that I spotted by accident just after I finally stood back up, curled tightly like a pug's tail with deep grooves running down their plump stems. Abundance beyond my wildest dreams.

I am beside myself with joy, swelling with newfound confidence. *I'm a real mushroom hunter now,* I think to myself. *I finally found my first morels.* I cradle them in my hands, lifting them to my eyes, a dream suddenly made true, inhaling their earthy perfume and committing it to my memory.

GATHERING EXERCISE

"STILL"

Today, choose a time and place to be absolutely still.
Do it with a plant, if you can
Make a circuit with your body and the plant, grounded in the earth
Even if that earth is just soil from a flowerpot
And the plant is your roommate.
Start at your feet and tell them to be still
Then your ankles
Your shins
Your knees
All the way up to the spinning brain between your ears
And tell yourself to *shhhhhhhhhhhh*
But don't forget to breathe.

Wildcrafted Sofrito

Makes 2 quarts

Sofrito is the base for all things savory in Puerto Rican cuisine. You can buy it in the frozen section of most well-stocked grocery stores, but homemade is undeniably superior to store-bought. It's also easy to make. Every Puerto Rican family has their own recipe, although every recipe has a few ingredients in common: sweet peppers, onions, lots of garlic, olive oil, and herbs like cilantro and recao (also called culantro). In my recipe, I like to include seasonal foraged elements that fit the flavor profile whenever possible. Wild onions of any kind are an obvious inclusion, but one ingredient that may surprise you is *Rudbeckia laciniata*, also known as sochan or cutleaf coneflower. The young leaves of the sochan have a slightly bitter, zesty, celery-like bite that adds a cheeky punch, pairing nicely with other herbs without overpowering them.

Hint: If you make this now and freeze it, you can use it later in the year when it's time to make another recipe!

½ pound aji dulce peppers
½ pound cubanelle or green bell peppers
1 cup lightly packed wild alliums (ramps, field garlic, or other wild onions work well, but you can substitute with chopped white onion)
1 cup whole garlic cloves
1 cup extra-virgin olive oil
1 bunch cilantro
1 bunch recao (culantro)
1 cup loosely packed chopped sochan
1 tablespoon kosher salt
½ teaspoon freshly ground black pepper
1 tablespoon finely chopped bee balm leaves
1 cup water

(continued)

Roughly chop the peppers and alliums, then throw them into a food processor with the garlic and olive oil. Pulse until the mixture is reduced in volume by about half and appears thick and chunky, like commercial jarred salsa. Stuff the cilantro, recao, and sochan into the food processor on top of the existing mixture along with the salt, pepper, bee balm leaves, and water. Pulse until it looks like a restaurant-style salsa—thinner and wetter, with a finely minced texture. The sofrito can be stored in the refrigerator for up to two weeks or in the freezer for up to six months (I freeze mine in ice cube trays and then transfer to freezer bags for easy portioning). Use as a marinade for proteins, a base flavoring for rice, or any recipe where you need a whole lotta flavor!

Curried Pheasant Back Vegan White Chili

Serves 6 to 8

Pheasant back mushrooms (*Cerioporus squamosus*) have a bad reputation for being tough and not worth the effort to collect, but when properly harvested and processed, they can be excellent for ushering in the beginning of spring mushroom season.

When gathering pheasant back mushrooms for the table, try to make sure that you select specimens with tight, dense pores. The pore surface should be a very light cream color, and the mushrooms should smell clean and fresh, reminiscent of cucumber or watermelon. Smaller specimens tend to be better for the table. Use your knife to slice off the tenderest portion of the mushroom, usually a few inches from the edge (this is sound advice for harvesting most polypores). I like to remove the pore surface by cutting or peeling it away, then brushing off any dirt or debris.

Pheasant back mushrooms may *smell* unusual when fresh, but they are actually quite meaty once cooked, with a mild flavor and a texture reminiscent of chicken breast. In addition to serving as a dead-on mimic for shredded chicken in this recipe, they also take beautifully to marinades with East Asian flavors, such as miso and soy.

½ to 1 pound pheasant back mushrooms
Salt and freshly ground black pepper
¼ cup olive oil
1 to 2 tablespoons chili powder
1 tablespoon garam masala
1 small bunch field garlic scapes or ramp leaves, chopped (use regular garlic or spring onion greens if you don't have these)
1 small yellow onion, diced
1 (15-ounce) can sweet corn, drained, or equivalent amount fresh/frozen
2 (15-ounce) cans white beans (any type), drained
1 (15-ounce) can full-fat coconut milk
2 cups vegetable broth
1 tablespoon soy sauce or coconut aminos
1 to 2 green chiles, diced (skip them if you can't handle the heat!)
1 to 2 cups full-fat coconut cream

Prepare the pheasant back mushrooms by removing the pore surface, then slice or tear them apart with two forks until the mushrooms resemble shredded chicken.

Toss into a cast-iron or other heavy-bottomed saucepan set over high heat with a pinch of salt and cook until they begin to sweat. Agitate the mushrooms in the pan until they have reduced in size by about half and have begun to form a crust.

Cut the heat to medium. Add the olive oil, chili powder, garam masala, garlic scapes, and onions. Cook, stirring occasionally, until fragrant (be very careful not to burn the spices). Add the corn, beans, coconut milk, broth, soy sauce, and chiles (if using). Bring to a simmer, then reduce the heat to low and simmer uncovered for 30 to 60 minutes, stirring occasionally, until vegetables are slightly softened.

Remove from the heat and stir in coconut cream one spoonful at a time until the chili reaches your desired thickness.

Serve with crusty bread or another equivalently delicious carb.

Fiddlehead Summer Rolls

Makes 8 rolls

Fiddlehead ferns are another short-lived seasonal vegetable, and one that causes significant confusion for new foragers. The term itself is confusing—nearly all ferns produce "fiddleheads," the curved shape reminiscent of the scroll on a violin, but only a few species produce *edible* fiddleheads. By far the largest and most delicious common edible fern is *Matteuccia struthiopteris* (you can read all about how to identify it by going to the Alphabetical List of Plants and Fungi in the back of this book). The curled-up shape always reminds me a little of a cooked shrimp, which is what gave me the idea to put them in these light, refreshing summer rolls.

Keep in mind that although ostrich fern fiddleheads are edible, they are *only* edible when young and still tightly furled, and they need to be cooked first.

For the dipping sauce:

1 tablespoon rice vinegar

1 tablespoon dark soy sauce

1 teaspoon fish sauce (optional)

Juice of 1 lime

½ cup peanut butter

½ teaspoon toasted sesame oil

For the wraps and filling:

1 cup fiddleheads

4 ounces dried cellophane noodles

½ cup lightly packed fresh Thai basil leaves

1 cup roughly chopped fresh greens (young dandelion greens and sochan leaves both work well)

8 to 16 magnolia tepals (optional)

¼ cup finely chopped alliums (ramp leaves or field garlic greens)

8 rice paper wrappers

To make the dipping sauce, in a small bowl, whisk together the vinegar, soy sauce, fish sauce, lime juice, peanut butter, and sesame oil until smooth. Set aside.

Prepare a medium bowl of ice water and set it beside your stovetop. Fill a saucepan with 2 inches of water and bring to a rolling boil, then add the fiddleheads. Blanch for 2 minutes, then remove with a slotted spoon and immediately transfer to the ice water. Set aside.

Place the dried cellophane noodles in a medium bowl and cover with cold water. Let soak for a few minutes, until the noodles are pliable and limp but not mushy. Drain and place in your workspace.

Drain the fiddleheads and place them in a bowl in your workspace. Prepare the Thai basil leaves, fresh greens, magnolia tepals, and alliums. You must work quickly when assembling the rolls, so make sure everything is easy to reach!

Pour about an inch of cold water into a cookie sheet or a low, flat baking dish and take out a clean cutting board to work on. To assemble a roll, dip one rice paper wrapper into the cookie sheet, allowing it to fully soak for a few seconds, until it feels gelatinous and floppy (this will happen very quickly). Remove it, lay it out flat on the cutting board, then place a little bit of each ingredient in the center of the sheet in a lengthwise fashion, similar to a sushi roll. Roll up and tuck in the ends, then repeat for each sheet.

Serve immediately with the peanut sauce for dipping.

Ramp Leaf Compound Butter

Compound butter is a 1-to-1 ratio recipe, so modify the amount of butter you use based on your ramp leaf harvest. While I've used ramps for this recipe, you can swap in any green allium leaf of your choosing (for example, garlic scapes, leeks, or spring onion). Consider it a sort of "master recipe" that can be used for any number of foraged treats. Get creative and experiment! Unlike recipes for pickling or fermenting that require precision with a scale, the ratios in this recipe can be visually estimated.

I use compound butter in a lot of my cooking, but the ones that *always* have me reaching for my allium compound butters are garlic bread, roasted squash, steak, and mashed potatoes (yum)!

Fresh ramp leaves, 1 part (by volume)
Unsalted butter, 1 part (by volume), softened just to room temperature
Flaky salt (like Maldon), to taste

Freeze a cookie sheet for at least 30 minutes. Prepare a sheet of parchment paper to fit on top of it.

Process your ramp leaves in a food processor on high speed until they resemble a paste. Add the butter and spin again on high speed until the butter is bright green and fully incorporated.

Scoop out the butter with a spatula and dump it into the center of your parchment-covered frozen cookie sheet. Spread into a single layer, leaving a few inches of parchment free at each of the edges. Add a generous sprinkle of flaky salt over the whole sheet of green butter, then roll it into a tight log, starting at one of the short ends of the parchment and rolling toward the other. As you roll, remember not to roll the parchment *inside* the butter. If you roll the parchment with the butter, you'll have a "cinnamon roll" of parchment that you will need to pick out from the butter roll every time you use it. Work rapidly to avoid melting the butter with the heat of your hands.

Twist the ends of the paper and cover with either beeswax wrap or plastic wrap to seal. Keep in the fridge for 3 to 4 weeks or in the freezer for up to 6 months.

Ramp Leaf Salt

Like the compound butter, ramp leaf salt is also a 1-to-1 ratio recipe. This recipe requires the ability to dehydrate, and because ramp leaves are usually gone by the time it gets hot, some sort of indoor dehydrator is recommended so nothing spoils. I use a Cosori oven-style six-tray dehydrator, and I find that it dries much faster and more evenly than the circular stacking-style dehydrators available on the market, but either kind will work.

Because blending, which will decrease the size of your salt grains, is a necessary step in this recipe, I highly recommend using coarse or flaked salt to avoid making your final product too fine. Ramp leaf salt adds a very attractive finishing touch to a dish with its vibrant green color, so you'll want to maintain the integrity of the grains. I typically buy coarse Diamond kosher salt, which holds up well to this process.

Fresh ramp leaves, 1 part (by volume)
Coarse or flaked salt, 1 part (by volume)
A desiccant packet (optional)

Process your ramp leaves in a food processor on high speed until they resemble a paste. Add the salt to the food processor and spin on low speed until the mixture is fully incorporated and resembles wet sand (if you spin it too fast, you may end up with salt powder when it dries).

Scoop out the salt paste with a spatula and spread it out evenly onto a dehydrator tray lined with parchment paper or a silicone dehydrator mat. Dry at 90°F for 6 to 8 hours, until no moisture remains.

Crumble the dried salt sheet with your hands or place the salt in a bag and crush gently with a rolling pin or a flat stone until your desired texture is achieved. Store in a jar in your spice cabinet and add a desiccant packet or shake occasionally to prevent clumping.

Lacto-Fermented Garlic Mustard Pesto

Makes 1 to 2 pints

Lacto-fermentation, simply put, is a process whereby *Lactobacillus* bacteria breaks down sugars in food and converts them into lactic acid. Sounds scary? It's not, I promise. Lacto-fermentation is probably one of the most predictable, safe, and beginner-friendly ways to preserve all kinds of fruits, veggies, and greens. It is easy to tell when a lacto-ferment has gone off (it will smell bad or have mold), and it can improve, deepen, and enrich simple flavors into something absolutely *wonderful*.

There are two basic kinds of fermentation: aerobic and anaerobic. Aerobic fermentation is commonly used in making products like kombucha and beer, and requires oxygen to support the yeast and transform sugars during the fermentation process. Anaerobic fermentation does not require oxygen, using enzymes to convert sugars into alcohols, gases, and acids. Lacto-fermentation is an anaerobic process.

To begin, you'll need a digital food scale that can weigh to the gram. These are fairly inexpensive (and depending on your lifestyle, you might already have one for weighing . . . other things).

To make the lacto-fermented garlic mustard:

A sterilized glass mason jar, or any glass jar with a sealable, airtight lid

2 cups fresh garlic mustard leaves

Non-iodized salt (This is very important! Iodine is antibacterial and may kill the useful bacteria involved in fermenting your food. I like using sea salt or pickling salt.)

A stone or fermentation weight (you can also use a zip-top plastic bag filled with water)

First, place your jar without the lid on your scale. Zero out the scale (usually done by hitting a button that says "zero" or "tare") and set the jar aside.

Fill up the jar with loosely packed garlic mustard leaves, making sure to leave room for your weight. Place the jar on the scale again and continue to slowly add garlic mustard leaves until you reach a nice, even number (like 200 grams, for example).

Now we have to do math, but it won't be too difficult! Take the final weight of your garlic mustard leaves (in grams), and multiply that number by 0.02 to find how many grams of salt to add. In this case, 2 percent of 200 is 4, so we'll add 4 grams of salt to the jar. Zero the scale out again to measure and add your calculated salt quantity.

Seal the jar and give it a good shake, making sure the salt coats the leaves. When they're evenly coated, remove the lid, place your fermentation weight on top of the garlic mustard leaves, and seal the jar up again.

Once per day for the next two days, remove the lid to let excess carbon dioxide escape. You will know your garlic mustard is ready when it smells funky, the volume has reduced significantly, and the jar suddenly has a lot of liquid in it that you don't remember adding.

From here, you can use your lacto-fermented garlic mustard to make your pesto!

To make the pesto:

½ to 1 cup lacto-fermented garlic mustard, plus 3 tablespoons lacto-fermentation liquid
½ cup grated Parmesan cheese
½ cup pine nuts or walnuts
¼ cup olive oil

Pulse the lacto-fermented garlic mustard and liquid, the cheese, and nuts together in a food processor to make a slightly chunky paste. Respin on medium speed and add the olive oil gradually while spinning, a drizzle at a time. The final texture should be soft and light but not overly smooth, like chunky almond butter but slightly runnier. I love adding a spoonful of garlic mustard pesto to sandwiches, pasta, and flatbreads.

Morel Risotto with Saffron, à la Debs

Serves 2 to 4

Morels are a highly anticipated seasonal treat, but it is important to make sure you cook them thoroughly before you dive in, as undercooked morels can make you very sick. One of my favorite ways to eat them is in risotto. I can remember a trip to Wisconsin with my friends Rudy and Debs, where, after a long week of camping, getting chewed up by mosquitos, picking mushrooms and ramp leaves, and searching for rare wildflowers to photograph, we finally arrived at a beautiful Airbnb yurt with a kitchen and a real shower. After cleaning ourselves up, we pulled out all our foraged treasures and made this meal. Debs insisted that saffron was *absolutely necessary* for the risotto, so Rudy searched high and low for saffron at all the local country shops until, miraculously, he found some. Debs was right. It was one of the best meals I've ever had, and we gorged ourselves on it for days. While saffron may seem like overkill, morels are special enough that I firmly believe it justifies the added expense.

Be warned: good risotto takes time and elbow grease. You can't set it and forget it, as it takes at least an hour of hands-on time to make the real thing. It's not a quick weeknight meal. You'll be stirring for what feels like a thousand years, and I know it's exhausting, but there are upsides: you can leverage your labor to make your partner or dinner guests feel guilty enough to take care of cleanup, and nobody else needs to know what *really* happened to the second half of that bottle of white wine. Plus, once you start spooning that lovely, creamy, cheesy, mushroomy rice into your mouth, you'll forget all about your sore biceps.

½ to 1 pound morels, chopped into bite-sized pieces

1 yellow onion, diced

¼ to ½ cup olive oil

Dry white wine (you'll need about half of a 750-ml bottle)

1 cup arborio rice (or a similar short-grain rice—I prefer sushi rice when I can get it)

1½ quarts hot broth (any type)

A generous pinch of saffron threads

3 tablespoons Ramp Leaf Compound Butter (page 76)
A generous serving of grated Parmesan cheese
Salt and freshly ground black pepper

In a heavy-bottomed pot, cook the morels and diced onion in the olive oil over medium-low heat until the onions are fragrant and translucent, 3 to 5 minutes. Deglaze with a glug of wine, then add the rice.

Reduce the heat to medium-low and add about 1 cup of the hot broth at a time, stirring constantly, making sure the rice grains are perpetually agitated. Don't add so much broth at one time that the rice begins to boil, or it will not cook properly. Agitating the rice grains aids in releasing the rice's starches, which gives the dish its creamy texture. Using hot broth during this process will significantly cut your cooking time, as it will absorb into the rice much faster without having to come up to temperature. You'll know that your broth has been absorbed into the rice when you don't see any more liquid in the pot and the grains begin to stick to the bottom of the pot. Every time the rice absorbs the broth, add another cup until you run out of broth. Keep stirring through the whole process and resist the temptation to raise the heat—if the heat is too high, it will cause the liquid to evaporate too fast, and your rice won't cook fully.

When all the broth has been fully absorbed by the rice, add the saffron threads. Pour in the wine a couple of glugs at a time until you have used half of the bottle, continuing to stir throughout the entire process. Taste the rice when about half the bottle of wine has been added. The rice grains should be creamy, fluffy, and tender, with no obvious crunchy bits. If needed, add more wine until the desired texture has been achieved.

Remove from the heat. Finish by folding in the ramp compound butter and cheese, making sure to lift the rice with your spoon to evenly distribute the butter and cheese throughout the pot. The rice should have a thick, porridge-like consistency, neither visibly wet nor dry. When served on a plate or in a pasta bowl, the risotto should hold its shape for only a few seconds before beginning to spread out. Season with salt and pepper. Serve hot.

To Be a Flower, Is Profound

Bloom—is Result—to meet a
Flower
And casually glance
Would cause one scarcely to
suspect
The minor Circumstance

Assisting in the Bright Affair
So intricately done
Then offered as a Butterfly
To the Meridian—

Emily Dickinson,
"Bloom—is Result—to meet a Flower"

APRIL 2022

Factory Coffee, Kalamazoo, Michigan

I'm looking out the open window of my favorite coffee shop at a flowering magnolia. Its petals (not even petals, really, but a type of primordial petal-like structure called a *tepal*) are beginning to shower the ground beneath it, blurry like a watercolor without my glasses on.

It's funny how things turn up in your life over and over again, landmarks in time. It's the beautiful thing about seasons, or perhaps the tragic thing, depending on what they remind you of. Before you even see the changes in the seasons, you smell them—the crisp air giving way to new grass, the faintest promise of florals, but mostly, to memory; a greening and warmth that caresses your spine with its vining fingers. To me, the smell of magnolia flowers is as good as a portal to the past. Time, which in the winter seemed to move at a snail's pace, now at once begins to spin on a scale I can perceive, every fresh morning bringing new changes to the landscape.

To the forager, scent fixes us where we are in time while at once reincarnating our own past experiences. The return of the magnolia flowers, or indeed any other scent, is nearly as good as a calendar, adding new layers to our own autobiographical memory like a tree growing new rings every year.

There are times where it simply feels too raw, too vulnerable to be outdoors in spring. As the season's many fresh smells come your way, there may be too many things to remember, and too many of them painful.

And so, I am retreating to the safety of my coffee cup today, shutting the window, unwilling to face the burden of the past as it wafts past me, invisible on the breeze.

APRIL 2002
The Pond, Pennsylvania

I'm a few dirty, bare footsteps behind Katie, both of us laughing as we race along the edge of The Pond. The stillness of winter and the shyness of early spring have given way to a rather more extroverted true spring, insects and frogs and birds chattering away as though they had never left, and the wisps of creeping green now erupting into a full carpet of stunning color.

"I hope they're still picking flowers together when they're fifteen," a soft grumbling voice muses, almost out of earshot. Katie and I look at each other, silently acknowledging that we both heard what he said. I nod and she nods back, indicating a solemn childhood oath.

Of course we will. We always will.

IF I WERE a perfumer, I like to think I would try to recreate my favorite places through scent, and first on my list would be The Pond. A tiny oasis in the Pocono Mountains, stewarded by Katie's grandparents, affectionately known to all of us as "Poppa" and "Tutti" (the latter so named because her oldest grandchild couldn't pronounce the German diminutive for grandmother, *Großmutti*), Poppa dug the pond out himself with a backhoe, taking advantage of the nearby creek and natural springs to feed it. He built a cabin on the property, complete with a porch swing fashioned out of a repurposed blue car seat with only a few springs poking out. (He would always argue that "it's only dangerous if you sit on that spot!")

Around the pond is nothing but acres of mixed pines and hardwoods, ancient lichenous boulders scattered like forgotten gods. The only way in or out is to locate an invisible dirt road off the main highway and let yourself in through a rusty gate. Do not try to find it, because you won't. In this place, my friends and I could be children, freed from the pressures of society. We were feral, catching bullfrogs and collecting frogspawn from the vernal pools, having mud fights and painting ourselves head to toe in sludge, swimming among the bluegill and pondweed, squealing in delighted surprise when the fish nibbled us, even as we waited, still and silent with our toes dipped in water, baiting their approach. We nibbled on tender young *Picea abies* spruce tips before we knew what they were, rolling the square needles between our fingers and sucking tacky resin off our skin.

The pond has no magnolia, but it does boast an abundance of florals. Blue irises in the heat of summer, bluets (which we swore were forget-me-nots), roses, violets, black cherry blossoms, water lilies, chicory, grape hyacinth, lily of the valley, and sprays of daffodils are flung far and wide, nearly as wild as the children who plucked them and fashioned them into bouquets. If I were a perfumer, maybe I could bottle that wonderful wildness, that smell of rich decay, of fungi knocking on the bark of old logs, ancient stone and fish scales, resinous pines and woodsmoke, mosses thick with musky secrets and florals dancing light and shimmering above them.

I am a forager, a collector of sorts, but I am not a perfumer. I feel the urge to return in the flesh, again and again, in search of that which will trigger my waking mind to recollect the dreams of my own past. If I could trap that memory, I would never apply it to my skin. Decades later, the feeling of that place lives deep in my pores, in the turning over of my skin cells, in my mind's eye, a feeling almost but not quite powerful enough to transport me.

Instead, I would keep it in a vial around my neck and open it only to catch the occasional whiff, and to remember.

MARCH 2020
Indiana

"We just don't know," my mother says through the phone. "I stopped by and found her in bed, she couldn't even respond to me. She's lost so much weight; she's just a shell. We don't know if she can live by herself anymore."

It's alarming to hear my mom sounding this fearful about *anything*. Julie Cerberville is the sort of person you might refer to as an "incredibly tough cookie," because she is. She has a tough, yet practical presence, much like her German ancestors, marked by an unflappable positivity, a permanent smile, and a fascinating ability to make friends everywhere she goes, a skill presumably learned during her upbringing as a military kid. Moving to a new state or country so often nurtures a special kind of resilience in children, one that people like my mother carry into their adulthoods. She barely *worries*. It always seems that no obstacle is too challenging or oppressive for her to work around. She reminds me of a dandelion. "You bloom where you're planted" is perhaps the best way of saying it, and it's what my mom has done for years.

Taraxacum **sp., Dandelion**

This had to be about my ninety-year-old Aunt Millie, whom my parents had recently brought to live in a nearby apartment after an Alzheimer's diagnosis.

Millie Herrera, a proper spitfire. Barely five feet two, maybe a hundred pounds dripping wet, she possesses an energy and personality that can be seen from space. She used to work out at the gym near her old house on Long Island every day, and I have countless memories of her bounding up and down the stairs like a teenager, salsa dancing on Christmas, and working the room at every family gathering with effortless charm, magnetic as a movie star.

It started with small, easily excusable things. She would repeat herself, call at odd hours. Easy to chalk up to her advanced age (after all, she was nearly ninety). Then it progressed—she began to wander, neglected to feed herself, grew anxious and reclusive. She was reluctant to leave her home, but unable to care for herself independently. My father handled the move, hoping to keep a closer eye on her and advocate better for her needs.

"We would only need a few weeks," my mother goes on. She hasn't asked me anything yet, but I know where this is going.

"Of course I'll do it," I respond, not giving myself time to think. "I'll come."

LESS THAN A week later, I'm driving east through Pennsylvania on my way to Milford. The air smells softer as I close in on the heart of the mountains, lightly floral and sweet with deep, wizened notes of hemlock and moss. The highway is uncomfortably empty, as though I'm driving through a post-apocalyptic America.

Every hour or so, I pass another digital road warning sign. AVOID GATHERINGS. STAY AT HOME.

I don't know much about this new virus, but it feels like it's closing in on me. One day, things felt normal, the next, this. We don't know how it spreads yet, but we know it kills, especially old people.

Old people like Millie.

The news cycle is consumed by speculation. I try to shut my ears and ignore the warning signs, but within days of my arrival in Pennsylvania, the world has effectively shut down.

I am terrified that any wrong move on my part could result in me being responsible for Millie's death. I try to maintain control of the situation as best I can and assume that any outside interaction could prove disastrous. With the various therapists, nurses, and social workers assigned to Millie's case, I institute nonnegotiable rules for the apartment. Nobody enters without a mask and gown. Everybody washes their hands immediately upon entry.

I bolt the windows and doors, lock the poisonous cloud I imagine beyond our threshold away. At night, I lie awake and worry about whether the air filtration in the apartment is shared with the rest of the units in the building.

MARCH UNFOLDS WITH the gut-turning suspense of a man on a wire. For the first spring in my life, I cannot go outside to greet the world, I cannot bring myself to open a window to smell the lichen, moss, and petrichor. The ecosystem seems unbalanced and confused, bewildered by the sudden absence of human activity and waiting for the other shoe to drop.

The red pines planted in a row along the park swell with life, pushing out thumbnail-sized pine cones like fat, green and pink jewels, thick with yeast blooms and oozing sap.

I can feel the crows gossiping about our species, speculating about why we've all gone into hiding.

I stare out the window and watch the dandelions grow in the unkempt lawn, untrampled by the typical recklessness of our feet.

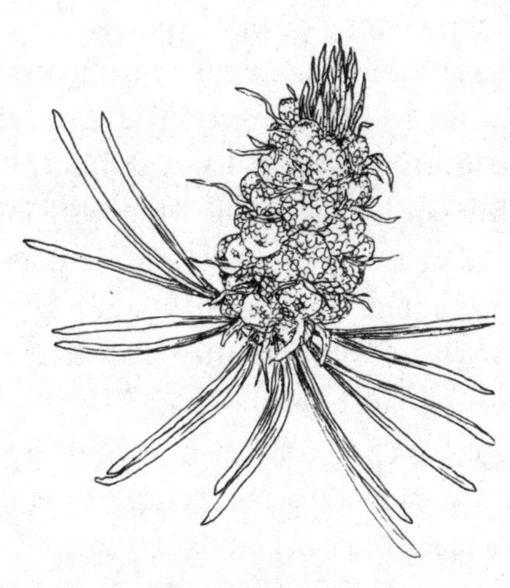

Pinus resinosa, Red Pine

APRIL 2021
The Micro Apartment, Kalamazoo

This apartment is my own. It is tiny, in a basement, and it has only one window, which looks out on an ambulance dispatch with a poorly directed spotlight that shines directly into my eyes at night. I have filled this apartment with objects that bring me joy—plants and rocks, paintings, fermentation experiments, jars of dried mushrooms, herbs hanging from the exposed pipes in my ceiling. My cats lounge on the windowsill in the sunshine, sometimes tackle and chase one another on the floor. It feels like home—more like home than anywhere I've ever been.

The virus has shuttered all of us away. Faces are reduced to eyes, outings are brief and spaced far apart. One year into my master's degree, and my colleagues and professors are little more than squares on a computer screen. I have never met them, but somehow I've still managed to meet all their pets.

I've grown comfortable with the long quiet and the empty streets, and I've found ways to make this crowded little place feel like home.

Every day, I walk to the river, where I visit with my plant friends and check on their progress. The peach tree has exploded in fuzzy pink blossoms, and the various flowering cherry and apple trees have donned outrageous white and pink and purple hairstyles. I collect some of the flowers to preserve in salt and sugar, to infuse into gummy candy and ferment into tea. Their season is brief, and within a few short days or weeks, the ground will be littered with spent petals.

The soundscape has shifted, too. Without me realizing it, the birds have returned to find mates and build nests for their children. Their chatter surrounds me, peppering through the trees. Their presence brings with it an assurance of the season, solidifying the return of green and growing things. Insects fill in the remaining auditory space with their buzzing oscillations.

The pandemic has made its mark on this place. I get the sense that I am entering into a nonhuman taskscape made far bolder by the retreating human activity over the past year.

My hands are rarely empty these days. I fill them with an ever-multiplying series of delights, regularly walking around town with a basket slung over my arm. Sweet cherry blossoms and magnolia tepals mingle in my basket, an unconsciously romantic arrangement of pink. I feel like a flower girl at a wedding between the sun and the earth as they celebrate their reunion.

Descending from the paved walkway into a secret pocket of unmarked forest near the river, I pluck young stinging nettles, *Urtica dioica*, growing from the rich soil with my bare hands, running my thumb and index finger gently upward on their hirsute stems to follow the direction of the stinging needle-like hairs, called *trichomes.* They can't hurt me if I grasp them in the direction that the hairs are angled. Nettles are all bark and no bite if you know how to work around trichomes; if you agitate them by crushing, drying, blanching, or even rolling them out with a pin, they'll be as tender and toothless as spinach, delicious and full of vitamins and minerals. Nettles also have a long history of use as medicine throughout many cultures. When I'm not hungry, I've been known to pick nettles and intentionally sting myself with them to address an aching knee or an annoying itch from a bug bite. The prickling pain is like a honeybee sting at first, rolling down your spine from the top of your head in a shuddering moment of overwhelming sensation, leaving behind raised bumps where it touched you. After a horrible moment, blood rushes to that area, and the histamine response takes over, providing a strange but enduring relief. The pain is worth it, as it so often is.

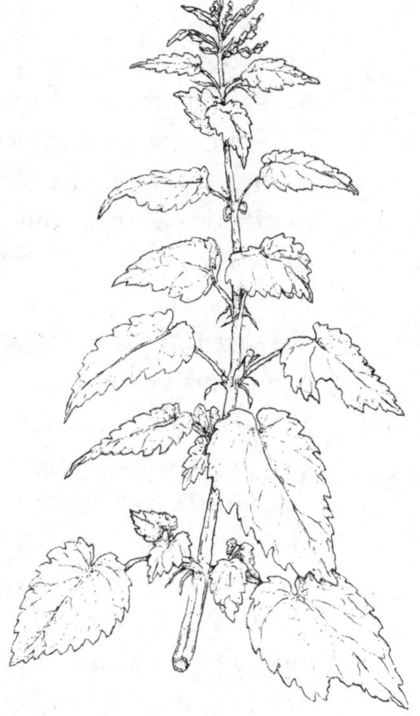

***Urtica dioica*, Stinging Nettle**

This spring has awakened something in me. Kalamazoo is undeniably rich in landscape. One can hardly walk down the sidewalk without running into some wild edible plant, whether in landscaping or in the vast green spaces that flow seamlessly through the city.

APRIL 2012

University Terrace, Butler University

The air is thick with the smell of botanical sex.

By which, of course, I mean that the fragrance of flowers is everywhere, dancing in the warm breeze. I'm feeling light, an excitement hedging on overwhelm settling on my shoulders. The world has yet to disappoint me too terribly, though it most certainly will.

The season is ramping up, always faster than I expect it to, and there is so much to do and see before the forest explodes into the anticipatory heat of summer.

The second semester of my sophomore year in college is coming to an end, marking the conclusion of a very unsatisfying, snowless winter and the start of a rainy early spring.

"Wonder what that tree is," my then-fiancé muses as we make our way up the hill from my apartment to the main campus. It isn't exactly one of those times when I think he's genuinely asking me a question, but an idle curiosity that happened to come out in the form of a question. Not one to miss out on an opportunity to play plant trivia, I pause, my eyes swiveling over to the tree in question, though my nose could have done the work of identifying it alone. I've known this tree for decades. Enormous, fragrant, cuplike blooms have sprouted, seemingly overnight, from the branches of a gnarled tree. "It's a magnolia," I answer, and then add, anticipating the next question about to exit his mouth. "It's edible, I think. The blooms of this kind taste a little like ginger, but I think some are supposed to taste better than others, if I remember correctly." I had learned to qualify myself with little hesitation marks, small linguistic uncertainties, so as not to seem like too much of a know-it-all. Not everyone likes it when I talk about plants.

I sense his hesitation, not entirely trusting my words.

"Better not risk it," he decides, dismissing the topic and continuing on up the hill in long strides.

I don't argue, but the inside of my mouth tastes acrid. I swallow, biting the inside of my cheek hard. The bitterness on my tongue is replaced by iron.

I wish I didn't keep saying "I think" when I really mean "*I know*."

I don't know why I do that.

As we ascend, I wade through wood nettles, sharp pain like shards of fiberglass stinging my bare ankles.

Wood nettle (*Laportea canadensis*), like stinging nettle, belongs to the same botanical family (*Urticaceae*), and like stinging nettle, it also sports painful hairs that can greatly reduce your personal enjoyment of having nerve endings. Its friendlier appearance, with larger leaves and softer leaf margins, is one of the dirtiest lies in all of botany. In fact, wood nettle hurts at least ten times *worse* than stinging nettle ever could. But in nature, prickles and thorns tend to be protecting something—often, something that we plant-eaters might like to snack on.

I want to run out of this place, inside, where it's safe and predictable. This air is sickening, and my legs hurt.

As we walk farther and farther from the magnolia tree, I chance a look behind and watch glorious pink showering the ground around it in a breeze-incited flurry.

Everything ends, sometimes even if it has only just begun.

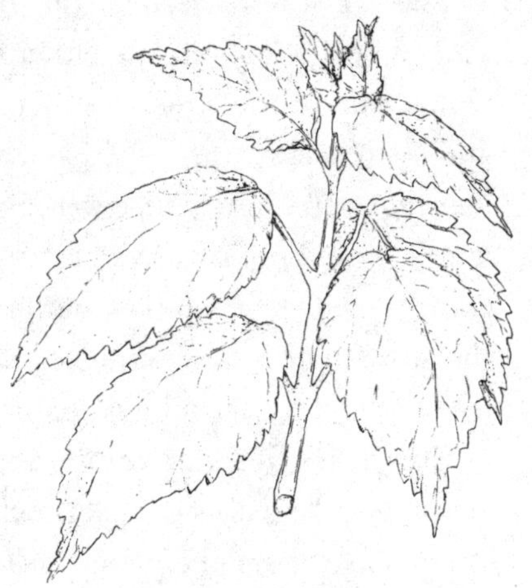

***Laportea canadensis*,**
Wood Nettle

APRIL 2017

Bag End, Indianapolis, Indiana

It looks like a hobbit hole," I gasped.

"Six hundred dollars a month, no utilities included," growled Greg, a scrawny man in his sixties who frequently looked over his shoulder at nothing at all. "I don't care if you have pets, just clean up after them. I don't wanna come in here and find shit on the hardwoods." He squinted suspiciously at me.

Noted.

To be honest, I was barely listening. I couldn't keep my eyes off the front door.

Looking back, the house should have been condemned. It was a poorly designed duplex at a bizarre angle. The basement flooded semi-annually, the kitchen was about three feet across and practically unusable, three of the four burners on the stovetop didn't light, the paint was leaded, every window let in a draft, and full conversations could be heard through the drywall separating us from our neighbor. It was rotting from its foundation, but all I could see was the coziness promised by that green front door with the rounded top and the knocker in the middle.

The house was nestled "on the corner of janky and swanky," in a neighborhood that was either terrifying or magnificent based on which side of the street you lived on. The backyard had a decaying and inexplicably unfinished fence, and a giant black walnut tree, *Juglans nigra*, that threatened to topple on the house every year when the winds picked up. Invasive plants littered the yard—English ivy, field garlic, daylilies, creeping Charlie. An overgrown lilac bushscrubbed the air with its clean fragrance, massive purple blooms swaying a few feet above the charming green door.

Everything needed cleaning and mending, but no matter how much I swept and scrubbed, it seemed to produce more dust and grime. The persistent smell of damp was impossible to rid from the bones of the house. Still, I loved it. It had a wildness, an unpredictability and charm to the peeling paint and tangle of weedy vines that consumed

Juglans nigra, Black Walnut

the backyard. I soaked English ivy leaves in water to make soap, plucked violets from the lawn and sucked the sweetness out of them, cut lilacs and brought them inside to perfume the air, purple and white and blue, watching as their short lives played out again and again on my dining room table, eventually covering it with a dusting of dead, brown flowers as they dropped from their branches.

Lilacs, *Syringa vulgaris*, are common, edible, and generally fleeting. There is a reason you don't often see lilac listed as a scent note in perfumes, despite being one of the most popular flower garden plantings. This is because the scent of lilac distills poorly, as it degrades quickly with or without heat. I have attempted many experiments to preserve it, including double extraction, ice cube distillation, and more, but only one method really works: enfleurage.

Enfleurage is an old perfumer's trick, and one you'll understand quickly if you store butter and garlic in the same container in the fridge: trap the volatile compounds using a fat and repeat with more plant matter until your fat of choice is extremely fragrant. Not all compounds are water soluble, and in the case of lilac, those responsible for scent are much better suited for extraction in a fat. You can achieve this for food purposes with butter or full-fat cream, but for skincare purposes you could use an oil that solidifies at room temperature (like coconut oil) or tallow. Simply place the flowers directly on or in your fat, let them infuse, remove the flowers, and replace with new ones until the fat smells the way you want it to. Keep in mind that once you add heat, the scent will disappear. Even as a perfume, the scent won't last, as the heat of your body will break the compounds down over time.

No matter how hard you try, nothing lasts forever. Not the flowers on the trees, not the trees themselves. When you try to make things stay the same, they'll rot from the inside.

Stasis is poisonous and, ultimately, impossible.

I was happy enough, for a while.

The lilacs lost their blooms, the grass grew taller, and I shut myself away from it, locking the green door behind me.

Time passed. I barely noticed, floating outside of my own body, a silent yet somewhat reluctant voyeur.

I FOUND A dining room table by a dumpster a block away. It was in poor shape, scratched and covered in white paint that had been stained a smoker-yellow over time. It came with six mismatched chairs, none of which seemed original to the table. I dragged it home and painted each of its four battered legs with winding trees, one for each season. The table sat unused in the dining room, a graveyard for junk mail and bills.

Across the street at the Presbyterian church, the giant saucer magnolia proudly donned her pink crown. *Come out and play,* she seemed to say, waving her branches at me. "I don't have time today," I pleaded, either coming or going. "Later," I promised.

I worked fifty hours a week as a veterinary office manager at breakneck pace, subsisting on caffeine, sugar, and cold French fries from the corner store. Saving lives and helping people was a rush, but my emotional resolve was beginning to fray. I was lonely, though I couldn't admit it. Every day was a revolving door of the same clients, the same problems I couldn't solve, the same smell of clinic disinfectant that wouldn't wash out of my hair. I was busy. I was tired. I had a partner, but somehow, I felt more alone than if I'd had no one. Keeping myself preoccupied with work and abandoning the plants and mushrooms for a while seemed to make things simpler. After a while, I stopped noticing them at all, letting them fade into a green haze while my working memory rusted over.

The ability to notice things is *everything* for foragers. Those who have honed a keen eye will spot fallen persimmons, oyster mushrooms sprouting from a sidewalk tree trunk, and grapes growing up the side of a chain-link fence; things other people don't even register. Like saprobic mushrooms, foragers deal in future waste and forgotten offerings, recycling and transforming things other people don't want or don't know about. We see the possibilities in the abundance surrounding us, and dream of new ways to turn water into wine. The gifts of plants and mushrooms are often given quietly, without glitz or glamour, but no less precious than gold.

***Pleurotus Ostreatus*, Oyster Mushrooms**

But not every gift that is offered is accepted. Later never seemed to come. The magnolia's flowers seemed forever doomed to rot in the gutter year after year.

APRIL 2022

Kalamazoo, Michigan

He's not here yet. Can you get here in 15 minutes? the text read. yup, I responded.

My layers of skirts are whipping every which way in the wind as I run to campus. I haven't had time to wash my hair or face, and I estimate that I'm on day three of the same mascara. Subconsciously, I can smell the flora in the air, but I haven't got the time to stop for it. I dash past plum-colored crabapple blossoms, weeping cherry, star magnolia, and redbud, and fling the double doors open wide.

Losa looks exhausted, and I know the constant anxiety is making her sick. We have been manning one of the concert halls together, running back and forth, trying to get ahead of problems, jumping in wherever we could. It's a big deal for the school, for us professionally. I'm trying to apply for PhD programs, and my future colleagues could be in these rooms. I feel it too—the over-caffeinated, stomach-curdling stress. I think my stomach acid might eat through my skin, and my eye twitch has come back in full force. The doctor says I have a vitamin D deficiency. I push it aside. *That's a tomorrow problem.* I pause. *Maybe next week's problem.*

"Did you get any sleep last night?" Lisa demands, concern emanating from her green eyes, laser-focused on mine. "Some," I offer, not willing to admit that I had, in fact, gotten no sleep at all and had only managed to relax after smoking enough weed to give Snoop Dogg the shakes. She's too smart for my evasions, but chooses to let this detail go.

"Did you eat?" she continues, this time her eyes narrowing. Lisa is tiny, but ferocious, and any hint that I might not be exercising the appropriate amount of self-care will result in a tough-love smackdown that will undoubtedly end with me in baffling, yet obviously overtired tears. She's always right.

"I've got an apple in my bag," I reply.

"It's not helping you there," she admonishes. "Take a granola bar. Your body needs food." She doesn't leave me an option, effortlessly turning on what my colleagues and I lovingly refer to as her "mom voice" and pushing the bar into my hands. I am in no position to refuse.

We work silently and frenetically, running cables to the stage and taping them down, adjusting projectors, and testing audio over the next hour or so. Eating has reminded my body of the severe caloric deficit that I've been running the past few days, and my muscles ache, my stomach grumbles. I silence it with the occasional long gulp of stale coffee, which I realize belongs to yesterday, but I don't care.

Satisfied with our progress, we collapse into the lecture hall seats. We sit like that for a moment, letting the chaos in the air settle around us. The quiet is eerie.

"Have you thought much about what you want to do after this?"

I haven't, not really.

I can't quite articulate it, but if I'm fully honest, I'm not quite ready to leave Lisa. Lisa's work is ingenious, emotional, and vivid, full of ideas that I can only begin to grasp. Her process is unlike that of anyone I've ever worked with before. Just being around her makes me more creative. When you watch her with her mind at work, you can practically see the electricity snapping and crackling around her pink curls.

Lisa understands me. She knows how my brain works, because hers works the same way. We're both touched with neurosis, rebels and contrarians at heart, yet always struggling with the impulse to burn our own candles at both ends to keep someone else warm.

I look over at her. She winces slightly. "Are you okay?" I ask.

"I think I'm just stressed," she says, then adds, "or it's an ulcer. I can't tell. This whole thing has been a lot."

Neither would surprise me. We've all been at it for days. If I smoked cigarettes, I'd probably be on the verge of eating them at this point.

"Would tea help?" I anticipate her refusal, adding, "I could probably use some myself, to be honest." My stomach churns unpleasantly, agreeing with my assessment.

She thinks for a second. "Actually, yes." She rummages through her bag and pulls out an enormous ring of keys. "One of these opens my office, and the hot water heater is by the filing cabinet. There should be tea somewhere."

"On it." I hop up, heading down the hallway to her office, when suddenly I stop.

Stomach pain. Tea.

Magnolia.

Magnolia *are* without exception, edible. I've known this for years, a decade before my certainty on the topic was ever questioned. There are over three hundred species around the world, and each has a different flavor profile and ideal uses. Some are deciduous, others evergreen, but misidentification is of only the mildest concern and can be avoided with even a passing knowledge of magnolia's botanical features. The huge white flowers of the giant southern magnolia, *Magnolia grandiflora*, possess moody notes of black tea, clove, and ginger, while the bright pink saucer magnolia, *Magnolia x soulangeana*, has a more straightforward, snappy ginger quality, ideal for pickling and garnishing sushi. Others, like the sweetbay magnolia, *Magnolia virginiana*, have a waxy white flower that smells deliciously of vanilla and warming spices. Several species of magnolia also have edible green fruits (more so when suspended in sugar syrup), a fact that I was delighted to learn after chasing a wild hare on the topic.

One of my favorite parts of any magnolia is the unopened, immature flower buds, still encased in a fuzzy velvet blanket to protect them from the uncertain temperatures. The flavor is often more delicate than their mature counterparts, lacking the tannic bitterness that can come with age. Fermented with salt, in sugar, or by themselves in a jar, the hope of a refill is sure to be a reason for your friends to return your mason jars to you. If you're in the habit of lending jars, you know how hard they are to get back! Magnolia buds, though unrelated to the ginger plant, can be similarly soothing to an upset stomach when turned into a tea (*yes, technically it's a tisane if it's not from the tea plant, but this is a champagne vs. sparkling wine situation, and I really just don't care about the distinction*).

I run outside and approach the same magnolia I breezed past earlier that morning. It is a star magnolia in half-bloom, white curls poking out from the ends of branches. I collect a few petals, which fall easily into my waiting hands, then a few of the fuzzy buds. I crush one between my thumb and forefinger, bringing it up to my nose to smell, then to my lips for a nibble.

Better not risk it, I hear that voice inside my head say.

I laugh at the thought, brushing it aside as the spicy warmth travels over the tip of my tongue. I suppose this is what it's like to trust your own mind.

The kettle sings. I crush and steep the magnolia in two chipped ceramic mugs, its spicy perfume amplified by the heat and rising through the steam to fill the cozy office. The comforting notes of ginger, clove, and black tea warm me from the outside in as my shoulders drop and my jaw unclenches.

Lisa accepts the steaming cup and sips, a look of surprise flashing across her face. "What flavor is this?" she asks, quizzically.

"It's magnolia," I respond confidently and without hesitation. "I just picked it in the courtyard."

"Oh!" she replies, smiling and taking another sip. "It's lovely."

APRIL 2020

Aunt Millie's Apartment

We've begun to take one masked walk outside per day, an hour or so after breakfast when others are unlikely to be out. I still imagine deadly viral particles floating in the air just by our heads, but I can't keep us indoors forever and we both need the exercise. It helps to regulate her, to tire her just enough that she can fall asleep at night. She hates the mask, doesn't understand it and pulls it off within seconds of it being put on, but we try anyway. Her therapist says it's okay, just make sure you wear one and don't walk near where others are. The walks are important, and even though she's losing cognitive ability, her body is still in excellent shape. Frankly, I need these walks, too, as my own mind begins to unravel.

Millie's vision is mostly gone in at least one eye, and severely impacted in the other. Not being able to see makes memory loss worse and more confusing. She clings to my hand and my arm, and we walk slowly, Millie in her oversized leopard-print coat (she always insists, no matter the weather), the pair of us shuffling down the road looking like one very slow, fluffy predator.

I point out the same trees and flowers every time, passing pieces of plants over to her to smell or taste as I tell her what they are. It feels almost normal, like any of the walks I can remember taking before the world shut down, just quieter.

"Dandelion. Have you ever had dandelion wine?"

"Lilac. Here, smell."

"Rose! It's prickly, be careful. I'll give you a little."

"Violet. This isn't a smelly one. Maybe on one of these walks we'll find some that are."

"The mullein is starting. Here, feel this leaf. Isn't it soft?"

My chatter seems to keep her entertained, though she rarely responds beyond a little vocalization here and there, usually in acknowledgment of something I've given her to smell. I've begun to identify a sort of simplistic lexicon in her syllables. She likes lilac, but not catmint. She likes the smell of magnolia, but not the taste. We always pause on the bridge over the creek just before we get back to the apartment. We both like listening to the water and the birds, and smelling the spring water tumbling over the rocks.

"What's your favorite flower, Aunt Millie?" I ask, leading her back in the direction of the apartment.

"I don't know," she responds. "What's yours, my darling?"

"Well, I'm not sure," I tell her. "I think it changes a lot." I pause, realizing that I'm giving her a cop-out answer. "Forget-me-nots. I think they've always been my favorite, even when I was little. I used to pick them at The Pond with Katie. I mean, at the time I thought they were forget-me-nots, but they were really another type of flower called a blewit. I like them both now. The one I thought was a forget-me-not, because of the memory, and the one that really is."

She thinks for a moment. "The little girl loved nature. Blueberries and little flowers up in the hair."

She doesn't remember who I am right now. That's okay. She remembers the little girl I was, a long time ago, and that's just as good.

"I think the little girl still loves nature. I saw her picking flowers and berries not long ago," I murmur.

"Oh yes, my darling," she says. "Many times. Of course she did."

We approach the smooth concrete slab that will lead us back to Aunt Millie's door. I spot a cluster of purple flowers to the left, nestled in the lawn.

"Maybe a nice salad with our sandwiches for lunch today, Aunt Millie?" I ask her.

"Oh yes, my darling. A nice salad," she echoes.

"I'll make a special one," I tell her.

"Wonderful, my darling."

I lead her back into the apartment and help her settle into the couch, turning on *The Golden Girls* to entertain her until lunch. "Thank you for letting me stay with you, Aunt Millie. I love getting to live with you," I tell her with a kiss on the cheek.

Letting her believe that she's helping *me* makes her feel good. It's easier for her to believe than the truth—that I'm actually here to care for her.

I'm realizing that after nearly a month of thanking her every day for letting me live with her, it's truer than I thought. My loneliness and grief had been consuming me back home. The alternative, even with the insanity that accompanies it, gives me a purpose I didn't know I needed.

This new, old place is a balm for my soul. With the sun warming my shoulders and the smell of green and flowers everywhere, things are finally beginning to right themselves. I am coming back into step with time, in a dance that feels like hope.

I'VE ALWAYS FOUND it a bit funny that women are so often likened to flowers in art, but even I have to admit that the metaphor holds at least some measure of universal truth. Not for perceptions of feminine fragility, fertility, seduction, or the bloom of youth, but for their determination; the strength to remain beautiful in the most inhospitable of conditions.

"I have grandchildren—you're my grandchildren," I remember Millie saying when I went off to college. "All three of you are my own grandchildren."

Flowers are an inherently sacrificial part of plants. They are fully expendable—organs designed only for attraction, then disposed of to make way for fruit, or in some cases for nothing at all.

Sometimes a flower is the point—an end in and of itself.

We sit for a while, watching idly as Rose, Blanche, and Dorothy concoct ever more outrageous ways to catch handsome (and hopefully rich) men. Millie hums the theme song. I go out and examine the purple flowers closer. I already know what they are—violets—but for the first time in many months, I feel a pull from them, an instinct to gather. The sweet scent hanging in the air around them is magnetic—I don't need to bury my nose in them to know that they are sweet violets, *Viola odorata*, but I do anyway for the pure joy of it, kneeling and pressing my face to the cluster, inhaling deeply.

Sweet violets are magic in the best way. Their five-petaled flowers can appear in purple or white, perched upon leafless stems like butterflies. A basal rosette of glossy, deep green, heart-shaped leaves completes each plant, paling in comparison to the flowers on the metric of looks alone but an equally delicious edible in their own right, mildly bright and crisp. The flowers are pH sensitive, containing anthocyanins, the natural flavonoids responsible for red, blue, and purple pigments present in many fruits, flowers, and vegetables. Extracting the color from a delicate ingredient like violets requires a precise touch. To make a homemade violet syrup, for instance, the violet flowers need to be trimmed to remove any green parts, then infused in filtered water that has been boiled with sugar to make a simple syrup, then cooled until it is hot, but not boiling. It is best to do this in a French press, which will allow you to hold the flowers under the level of the liquid for maximum extraction. After infusing for several hours, you will strain your flowers out and be left with a sapphire blue sugary liquid smelling strongly of sweet violets, which can be altered to your preferred shade by either adding an acid or a base. A couple drops of lemon juice or vinegar will turn it amethyst purple, a sprinkling of baking soda will turn it emerald green.

I pluck a few, catching them between my fingers and collecting them in an empty mug. They fill my nose with a soft nostalgia.

Flowers test our ability to love. They captivate and delight us with their beauty and fragrance as they bud and blossom, and just as quickly, their initial charms fade away.

I bring the mug in to put it in the fridge until I'm ready to prepare lunch, the vibrant fragrance spilling into the room like a beam of sunlight. I pause by the sectional, where Millie continues to watch her show. "Smell, Aunt Millie," I say, holding the cup under her nose, "they're violets. Aren't they lovely?"

She smells, and an odd look brightens her face. "Oh, yes!" she exclaims. "This smells just like . . ." She stops, obviously a bit agitated, and starts rubbing her hands together. ". . . the sink."

"Of course it does," I realize, a memory suddenly returning to me. "The little soaps in the jar." They were purple and pink, shaped like roses and seashells. I always thought they looked like candy.

"Did you know you can eat them?" I ask. "Not the soap, but the flowers that the fragrance comes from?"

"Yes, very nice. Wonderful," she says agreeably. I've grown accustomed to our conversations ending this way. It doesn't bother me anymore. I no longer feel that desperate need to keep her grounded in my reality.

"I think I'll put these in our salad today," I tell her, squeezing her shoulder.

For the first time in weeks, my head begins to swirl with inspiration. Should I make soap? Syrup? Could I laminate violets between sheets of pasta dough to preserve their color? Ferment them into vinegar or make candy, holding on to their charming scent for just a little longer?

To love a flower is to watch it change. Nothing can stay the same forever; even if you preserve it you will have to accept that it will change.

If I were a perfumer, I don't know that this smell, this memory, would fit in a bottle. I don't think I'd need to distill it, though the thought tempts me. Instead, I would choose to simply return to the violets, year after year, clinging to the promise of their return like a life raft. Their re-emergence every spring is, thus far, an oath that remains unbroken.

How sweet it must be to be loved through the profound strangeness of being human, from beginning to end.

MAY 2022

Bells Brewery, Kalamazoo, Michigan

Lisa, Kennedy, Justin, and I sit around a table in the sun, neglected flagons of beer and cider perspiring in front of us. I can recall our laughter—broad and uninhibited, bright and free. The vivid pink and green remainders of a magnolia and nettle tiramisu sit in a casserole dish at the center, torn apart by my joyful, hungry, carefree friends.

Something new, yet impossibly old blooms in me.

I must continue to search for it, tracking it down until I recognize that forgotten fragrance, trusting on my memory to lead me back home to who I am again.

GATHERING EXERCISE

"GIFTS"

Find a flower outside. It may be one planted in a garden, or it may be one growing of its own free will in a place that nobody chose for it.

Ask the flower's permission to pluck it, and listen for the answer.

If the flower consents, thank it for its sacrifice.

Bring the flower indoors and place it in a vase or a glass of water.

Watch over the flower and love it until it dies.

Fermented Magnolia Bud Tea

Makes 1 pint

Magnolia has both medicinal and culinary uses that can be traced back to several Asian countries, in particular China, Japan, and Korea, where it is sometimes employed to treat stomach upset and anxiety as well as to flavor teas and rice. While pickling is a popular choice for magnolia, I generally prefer to lightly ferment and then dry the unopened buds for tea, as fermentation magnifies the spicy complexities and strength of the unique flavor compounds. Magnolia buds in many cases can be harvested twice per year, once in spring and then again in the fall.

For this recipe, you will want to gather unopened magnolia buds no larger than two inches long and still completely covered in fuzz. In my experience, the buds of *Magnolia x soulangeana*, the saucer magnolia, work best for this recipe, because the flavors tend to be strong and pronounced in the budding stage without the soapy or bitter flavors that may be present in some other species. Always taste your buds before collecting a lot of them! You won't need to add anything to this ferment, as the wild yeast present on the buds will do the work of digesting some of the plant material, enhancing the natural flavors.

½ pint fresh magnolia buds
1 pint-sized jar
A desiccant packet

Place the magnolia buds in the jar and screw the lid down finger-tight. Leave in a warm, dark place for 3 to 5 days, burping once per day. When the interior of the jar appears very humid and the scent of the buds is strong, funky, and perfume-like, remove the lid and dump out onto a dehydrator tray, spreading the buds out so that they are not touching one another. Dehydrate at 105°F for 8 to 12 hours, or until the buds can be easily crumbled between your fingers, with no moisture remaining.

To make tea, infuse two buds in a mug of hot water for at least 4 to 6 minutes. To save for later, let cool and then store in a clean, dry mason jar with a desiccant packet for up to 6 months. If you don't have a desiccant packet, shake well to reincorporate before using.

Nettle and Magnolia Tiramisu

Serves 3 to 6

The matcha-like sweetness and grassiness of nettles and the spicy ginger flavor of magnolia flowers marry together beautifully in this twist on a classic dessert. It can be made ahead of time and stored in the refrigerator, plus it makes an impressive finishing touch for a dinner party. I always reserve a small amount in a container for my midnight munchies, since I never seem to have any leftovers after serving it to others. No coffee required!

8 ounces mascarpone cheese
1 pint heavy whipping cream
⅓ cup powdered sugar
1 teaspoon vanilla extract
1 cup fresh nettles (or ½ cup dried)
1 (7-ounce) package ladyfingers
2 cups lightly packed fresh plucked magnolia tepals

Using a hand mixer in a medium bowl or in a stand mixer, beat the mascarpone cheese, cream, powdered sugar, and vanilla together on high speed until stiff peaks form, 1 to 2 minutes. Place in the refrigerator to chill.

Meanwhile, brew a strong tea using either fresh or dried nettles in about 2 cups of hot water. When the tea is a deep green color, discard the nettles and pour the tea into a shallow bowl to cool. One by one, dip the ladyfingers in the cooled tea and coat fully (try not to drench them or they will get too soggy and break).

Place a layer of ladyfingers along the bottom of a 9 x 13-inch baking dish. Next, add a layer of your chilled cream mixture, smoothing over with a spatula, and then sprinkle a layer of magnolia tepals on top. Repeat these steps, ending with magnolia tepals for a total of two layers.

Serve immediately or store in the refrigerator for up to 48 hours.

Violet Jelly Milk Tea with Violet Syrup

Violet is a fleeting ingredient, only available for a few weeks in the spring, but it lends itself well to sweet applications like jellies, candies, and syrups. This violet jelly milk tea is something that I look forward to enjoying each spring, and it always reminds me of my aunt Millie. It is important to note that violets contain pH-sensitive compounds called *anthocyanins*, and if it matters to you, the purple and blue colors will be altered by using tap water, which has an unreliable pH. Use distilled water (which will have a neutral pH of 7) to preserve the color, then add lemon or lime juice one drop at a time to turn it a lovely amethyst purple!

For the syrup:
2 cups granulated sugar
About 2 cups distilled water
1 cup violets, preferably scented, with all green parts removed
Lemon or lime juice (optional)

For the jelly:
7 grams gelatin powder (about 1 package)
3 tablespoons cold water (tap is fine)
1¾ cups violet syrup, warmed

For the milk tea:
Milk of choice, to taste
Ice, for serving
Brewed black tea, cooled
Violet syrup, to taste

To make the syrup:
In a medium saucepan, combine the sugar and distilled water, bring to a boil over medium-high heat, and boil for about 6 minutes, until the sugar fully dissolves and liquid begins to thicken. Set aside to cool until

(continued)

the mixture reaches about 180°F. Place the violets in a French press and pour the sugar mixture over the top. Place the lid of the French press onto the carafe and press down on the plunger slightly until the flowers are fully submerged. Steep for several hours, until the water is a deep sapphire blue and the petals appear translucent, about 1 hour. Slowly push the plunger all the way down and strain the liquid into a jar, discarding the petals. If desired, add a few drops of lemon or lime juice to turn the liquid purple. Set aside 1¾ cups of warm syrup for the jelly and bottle and store the rest in the refrigerator to use for up to a month.

To make the jelly:

In a small bowl, mix the gelatin granules with the cold water and let dissolve. Add the gelatin mixture to the warm violet syrup and stir. Before the gelatin can stiffen too much, pour into the mold of your choice (ice cube trays also work fine for this). Let the jellies rest for at least 2 hours, then remove from the molds when fully set.

To make the milk tea:

Add several jellies to a glass with milk (I like oat milk), ice, and a splash of black tea. Top it off with a bit of violet syrup for an extra hit of floral goodness. This drink is excellent when paired with a flaky, sweet pastry, such as a Danish or croissant.

Nettle Cake with Lilac Frosting

Serves 8

Lilac has a reputation for being a bit of a diva. It's true—it can be tricky to work with, and nearly every preparation will destroy the fragile aromatics that make it so distinctive. The only thing I've ever found that works is cold enfleurage—trapping the aromatics in fat molecules, which encases them and preserves them for a while (usually this is done in tallow or lard). Usually, enfleurage is thought of only as a perfumer's trick, but a savvy forager learns from everyone! This recipe uses butter for our fat source, but it should be noted that, even with this method, the aromatics still won't last very long, perhaps two or three days with good storage. Enjoy them while you can, and eat this up, as my Southern neighbors say, "right quick!"

For the lilac frosting:

1 cup (2 sticks) unsalted butter, softened (this will get refrigerated, but it needs to be spreadable in the first step)

2 heads lilac flowers, removed from stems

2 cups powdered sugar

1 teaspoon vanilla extract

¼ to ½ cup milk or cream (for thinning)

1 to 3 tablespoons blackberry or elderberry juice to approximate the lilac color, or a natural food coloring such as beet powder (optional)

For the cake:

½ cup (1 stick) unsalted butter, plus more for the pans

1½ cups granulated sugar

4 eggs

2 cups all-purpose flour

2 teaspoons baking powder

Hearty pinch of salt

¼ cup dried, powdered nettles

1 cup milk

¼ cup vegetable oil

1 tablespoon vanilla extract

(continued)

To make the lilac frosting:

Spread the softened butter in an even layer in a dish with a lid (a glass Tupperware-style container works well for this). Press a layer of lilac flowers into the butter and place a sheet of plastic wrap down over the top. Put the lid on over this and let it chill in the refrigerator for 2 hours. After 2 hours, remove the flowers, discard them, and replace them with another layer of fresh flowers. Repeat this at least four times. Each time you add new flowers, you're also giving the butter fresh molecules to "catch," which will create a stronger lilac flavor.

When you decide the butter smells strongly enough of lilac, remove it from the refrigerator and let it come to room temperature. Keep the lid on.

When the butter is soft, add the powdered sugar, vanilla, and milk to a bowl or a stand mixer and beat on medium-high speed until stiff peaks form to make your frosting. Add the juice or food coloring, if using. Store in a container with a lid in the refrigerator and keep tightly closed until ready to use.

To make the cake:

Grease two 8-inch round cake pans and preheat the oven to 350°F.

Combine the butter and sugar in the bowl of a stand mixer and cream together, then beat in the eggs. In a separate bowl, sift the flour, baking powder, salt, and powdered nettles. Add the butter, vanilla, and egg mixture to the bowl with the dry ingredients and stir to combine. Slowly add the milk and vegetable oil while stirring until the batter reaches a silky consistency.

Divide the batter between the prepared pans and bake for 25 to 30 minutes, until a toothpick can be inserted and removed cleanly. The top should be slightly puffed and gently crisped at the edges.

Remove the cakes from the oven, carefully remove from pans, and cool on wire racks fully. Frost the cakes, adding a generous amount of frosting between each layer, then decorate with more lilac flowers and dig in immediately to enjoy the fleeting loveliness of lilac before it drifts away into the ether.

Wildflower Laminated Pappardelle

Makes about 2 pounds pasta

Laminated pastas are a delightful project for the artistic cook, plus they don't *require* any special equipment aside from a rolling pin (a pasta machine is nice but not necessary). They're a great way to enjoy the flavors and colors of the flowers you forage, and they can be preserved for use throughout the year. You'd be surprised at how many flowers are edible, and in a pasta like this, colorful ones like violets, petunias, dandelions, redbuds, and forget-me-nots work particularly well. I love making laminated pastas that preserve the colors and flavors of each season and serving them with seasonal sauces and pestos (*psst, this is a great way to try out your lacto-fermented garlic mustard pesto*)!

A note on flour: Try not to use all-purpose flour for this recipe. All-purpose flour has a lower gluten content and a larger grind size, and will make a stiffer, less pliable dough that can't always be stretched out thinly enough to see all the pretty flowers pressed between the layers. If you can't find "00" flour (usually it's in the pasta section at the grocery store, and sometimes it's called "pizza flour"), I strongly recommend using bread flour, since it is milled more finely and will be better to work with.

---

2 cups "00" flour (or similar fine-milled flour)
3 to 4 egg yolks
Pinch of salt
½ to 1 cup edible wildflower petals (you can also add fresh herbs like basil, thyme, or oregano for color and flavor)

Pour the flour into a mound onto a cutting board or clean work surface. Create a well in the center, then add the egg yolks and salt. Knead together with clean hands until you have a springy dough that holds together and has a smooth, somewhat yielding, hydrated texture. Cut the dough into four equal parts with a bench scraper or a knife, then

(continued)

roll each into a ball and wrap semi-tightly in damp tea towels. Let rest for 1 hour.

If you have a standard pasta machine, roll the dough through on setting 6 until you have thin sheets of dough. Lay them beside one another on a floured surface. Lightly place flower petals and herbs on top of one sheet and stack another layer on top, sandwiching the flowers and/or herbs between the two pasta sheets. Roll through your machine until you can see the petals through the dough.

If you do not have a pasta machine, use a rolling pin or a clean, empty wine bottle to roll one ball out onto a floured surface until you have a thin sheet, but not so thin that it begins to tear when you handle it. Working quickly, arrange your flower petals and herbs all over the surface, then cover with a slightly damp tea towel to prevent it from drying out. Roll out a second ball, remove the towel from the first sheet of dough, and place the second sheet over the first. Using your rolling pin, roll the stacked sheets out as thin as possible until you can see the petals through the dough, then cut into wide noodles and hang on a rack to dry. I hang my pasta on a collapsible laundry drying rack, but you can also use coat hangers, baker's racks, or even the back of a chair. Once they have dried completely and can be snapped in two without bending, you can preserve the pasta in sealed jars for later, or you can use them right away without drying (just coat them in a bit of flour so they don't stick together when you boil them). The pasta will be shelf-stable in airtight jars for up to 6 months, and can be stored indefinitely in the freezer in zip-top bags.

Mugolio

Originally invented by Italian forager Eleonora Cunaccia in Trento, Italy, mugolio is a fermented syrup traditionally made from the unripe cones of the mugo pine (*Pinus mugo*), though many different conifers are amenable to the process. Mugolio is straightforward to make, an easy 1-to-1 ratio recipe by volume, which can be eyeballed rather than meticulously measured, but requires a firm measure of patience, as it won't be ready to enjoy for at least two months. Make it at least six months ahead if you want the richest, deepest flavor. My favorite conifers to use here in the United States are the pinyon pine (*Pinus monophylla*) and the red pine (*Pinus resinosa*), but really any spruce tree will work, plus spruce can be gathered a bit later in the season just in case you, like me, are a forgetful neurodivergent. Juniper cones also work in this process, as do green cedar cones. Learn to identify and stay away from toxic conifers such as yew, which can be fatal.

You'll notice that I specify "unwashed" conifer cones. This is because we want to preserve the colonies of yeast that are stuck to the resin on the cones, as they will be responsible for fermenting our mugolio. You can give them a light rinse and pick off any questionable bits, but don't scrub!

Throughout the process, be aware of any questionable smells or growths. Mugolio has a very low pH and is immune to many foodborne toxins, such as botulism, but it can still grow a variety of unpleasant molds and yeast colonies. Using a fermentation weight to press the pine cones underneath the level of the liquid will address most problems in advance, but if you see any mold starting to grow, throw it out and start over. There is almost always some kind of conifer in season that can be used for making mugolio. If using pine or spruce, your cones should be no larger than your thumb and still appear green and moist inside when split open. Older cones simply won't work at all.

Avoid using white sugar, as the end result will have an unappetizing mucus-y color and somewhat dull flavor profile. You need some of the sugar's natural molasses to achieve the depth of flavor and richness of color that mugolio is known for, so I recommend using raw, brown, or turbinado sugar. It should be noted that honey can be used in this recipe

(continued)

in place of the sugar at the same ratio, but the end result will be more of an infused honey than a true fermented mugolio. Still tasty and well worth making, just not quite the same. Other sugar substitutes such as xylitol and stevia aren't recommended, as the wild yeast present on the cones doesn't prefer to eat their sugar in that form.

1 part raw, brown, or turbinado sugar (or honey)
1 part unripe, unwashed pine cones

If you are using sugar, place a single layer of cones along the bottom of a clean mason jar. Add about an equal amount of sugar to cover, packing down lightly with a spoon. Continue layering sugar and cones in this manner until you reach about an inch of headspace from the top, ensuring the top layer is sugar.

If you are using honey, simply place all your cones in the jar at once and pour the honey on top, letting it settle in the space between the cones.

Place a fermentation weight inside the jar to hold down any cones that may float to the top during the fermentation process, seal the jar, and place it on a sunny ledge.

By day 2, you will need to burp the jar (release the carbon dioxide by cracking the lid open, then closing it again). If you are using sugar, it will look wet—this is because the sugar is drawing the water out of the cones. Burp once per day for 7 days, then remove the jar from the windowsill and store in a cool, dark place (such as a closet) for a minimum of 2 months, preferably 6. Continue burping once per week for at least 3 to 4 weeks, or more if need be. Inspect regularly for visible mold or any bad smells.

At the end of the fermentation period, strain the liquid into a pot and discard the cones, then bring the liquid to a soft boil to dissolve any stubborn sugar crystals. Stir and add water to thin or more sugar to thicken as needed until the mugolio reaches a consistency similar to that of molasses. Store in airtight mason jars at room temperature for up to a year. Serve drizzled on soft cheese, poured over vanilla ice cream, or used to flavor sodas and make mixed drinks, etc.

Spruce Tip Granita

Serves 2 to 4

Spruce tips are a short-lived seasonal treat, the tender new growth found on the end of spruce branches, much lighter in color than the more mature needles from previous years. Different species can have wildly different flavors; some taste like lime and grapefruit and fresh herbs, while others taste disappointingly of chalk. I always recommend taste-testing tips from different trees before gathering a whole basketful from a tree you may not enjoy. Always be careful not to collect *all* the tips from a given branch. Spread out your harvesting and make sure you are mindful of the tree's needs so it can continue to grow and provide you with food for years to come.

If you've never had a granita before, they're pretty similar to the chunky shaved-ice treats loaded with red and blue sugar syrup that they sell on boardwalks on hot summer days, except granita is even easier to make. This refreshing treat doesn't require an ice cream maker or any special equipment, plus it plays up the bright, citrusy flavor of the spruce tips without being too sweet or cloying.

1½ cups water
½ cup granulated sugar
¼ cup freshly squeezed lime juice
¼ cup loosely packed spruce tips
Zest of 1 lime

Combine the water, sugar, lime juice, and spruce tips in a blender and blend until smooth. Transfer the mixture to a shallow glass baking dish with a lid, such as Pyrex, and place in the freezer. The larger the container, the faster this process will go, as more of the mixture's surface area will be exposed to the freezer.

Every 30 minutes for 2 to 3 hours, use a fork to scrape down the sides and bottom, making sure to stir thoroughly. Ideally, you want

(continued)

large ice crystals to form. When the mixture is no longer loose, but not rock-hard, a slightly chunkier texture than a snow cone, serve immediately in bowls with a sprinkle of lime zest.

If you overdo it in the freezer and accidentally let it get too hard, don't panic! You can save it. Take it out of the freezer and let it thaw until you can break the pieces up, then puree it in a blender or food processor with a splash of water for a more "sorbet" type of dessert.

Rose Sugar

Makes 1 pint

When making both sugars and salts, I usually prefer working with fresh ingredients. Traditionally, many people who make herbal or floral sugars recommend blending your sugar with dried versions of plants to avoid spoilage issues. However, I find that doing so creates an inconsistent blend that regularly needs to be recombined by shaking, since the plant material is lighter than the sugar and will separate into its own layer. Instead, I like to reverse the order of things—start with a fresh product, add sugar, then dry the whole mixture together. This will still be a totally shelf-stable product, but it won't separate. Sugar also has hydrophilic (aka "water-loving") properties, which in this case means it will happily "suck in" the flavor compounds of the fresh rose petals through the water they contain when the two ingredients are agitated together.

When selecting your rose petals, be mindful of your source. Cut roses are often treated with antifungals to make them last longer, and they may be harmful for you to ingest. All rose petals are edible, but my favorite to use for sugar are those of *Rosa rugosa*, the invasive beach rose. I find them growing commonly in landscaping, especially near parks and malls.

2 cups lightly packed rose petals (choose pink or red petals if you want colorful sugar!)
1 cup granulated sugar
A desiccant packet (optional)

Blend the roses into a paste (work quickly to avoid browning from oxidation), transfer to a bowl, and mix with the sugar until the sugar resembles wet sand. Spread out onto a cookie sheet or dehydrator tray and dry in the sun or in a dehydrator at 100°F for 4 to 6 hours, until dry. The sugar combined with the low temperature in a dehydrator will prevent the rose petals from oxidizing.

Place the dried sugar in a bag and use your least favorite textbook to break up the clumps until granulated. To preserve fragrance and flavor, keep in a tightly sealed jar with a desiccant packet. It will maintain both its scent and color for at least 6 months if kept out of the sun in a cool room.

Of Gold and Red

A wee child toddling in a wonder world,
I prefer to their dogma my excursions into
the natural gardens
where the voice of the Great Spirit is
heard in the twittering of birds,
the rippling of mighty waters,
and the sweet breathing of flowers.

Zitkála-Šá (Gertrude Simmons Bonnin),
"The Great Spirit"

MAY 2021

Indianapolis, Indiana

It is said that if you, by some miracle, are able to usurp the geometry of light and travel to the end of a rainbow, a pot of gold will be there waiting for you.

Of course, this might be nothing more than mythology, intended only to entertain children and sell cereal, but in a sense, or at least, if you're a mushroom hunter, this is actually, absolutely, 100 percent fact.

June is approaching within the week, and I've just crossed the northern border of Indiana on my way to Indianapolis. I finally left my job at the veterinary clinic in 2020, burnt out by overwork and compassion fatigue, leaving me with a profound tic in my left eye that just wouldn't quit, no matter how diligently I cut back on my caffeine habit. The pandemic had worn down on me, forcing me to spend too much time in my own head. Within that solitude, a few things became clear: first, that I needed to move on with my life, and second, that I absolutely couldn't do that in Indiana. I spent the majority of my twenties here, going to school, working odd jobs to make ends meet, waiting around every year, secretly hoping that the next one would be when my life could start. It was time for a fresh start, and for me, that looked like grad school up in Michigan. The air seemed lighter up there, not smoggy with my past routines.

It feels bizarre to be back in a place that made me feel so insignificant after leaving, so utterly stuck, and it is especially confusing to be returning of my own volition.

All I see in every direction are cornfields of wasted time and broken promises.

I still have friends here, of course. Best friends, in fact. Being around them helps, makes me feel less on edge. But this *place* . . . I don't like how I feel when I'm here.

There are ghosts all around me.

The worst part is that most of them *are* me.

The only reason I've agreed to return is because a local arts organization has invited me back to display some work in one of their galleries for a monthlong show. I'm grateful, honored to be asked at all. The back of my car is a cacophony of framed scores, sculptures, boxes of electronics waiting to be installed. Some of them were written here, others in Michigan, reflective of several years of dramatic personal development.

I-69 seems to go on for an eternity, and I can feel the sun beating down through the windshield as I grip the steering wheel and grit my teeth.

My favorite foraging basket, a green one from the food co-op near my apartment in Kalamazoo, sits on the seat beside me, bouncing with the Honda along the ragged highway. Having it there is comforting, a concrete reminder of the choices I've been able to make since I left this place. The life I've been able to build. The *newness* of it all.

My dashboard lights up. *Damn.* I need gas. Time to start scanning the upcoming exits for the nearest station. I'm a little anxious—finding a place to fill up in corn country isn't always easy.

A few miles pass as I hold my breath and try to coast as much as possible. Fortunately, a large sign promises my choice of gas options just off the highway. Additionally, I take note of a familiar brown sign, indicating a nearby woodland area just a few miles off the same exit.

I check the time. I'm not due in Indianapolis for at least a few more hours.

Fuck it. The city can wait.

The hunt is on.

The summer breeze wafts lazily through a bobbing, unkempt patch of wild carrot, *Daucus carota*, as I park and begin to walk toward the trail. Within a few hundred feet, the trail morphs into a boardwalk that cuts through a bog, filled with the overpowering, warm stink of standing water and muck. Croaking, buzzing, splashing, chittering sounds come from all directions, and redwing blackbirds race one another through the cattails, leaving explosions of sulfur-yellow powder in their wake.

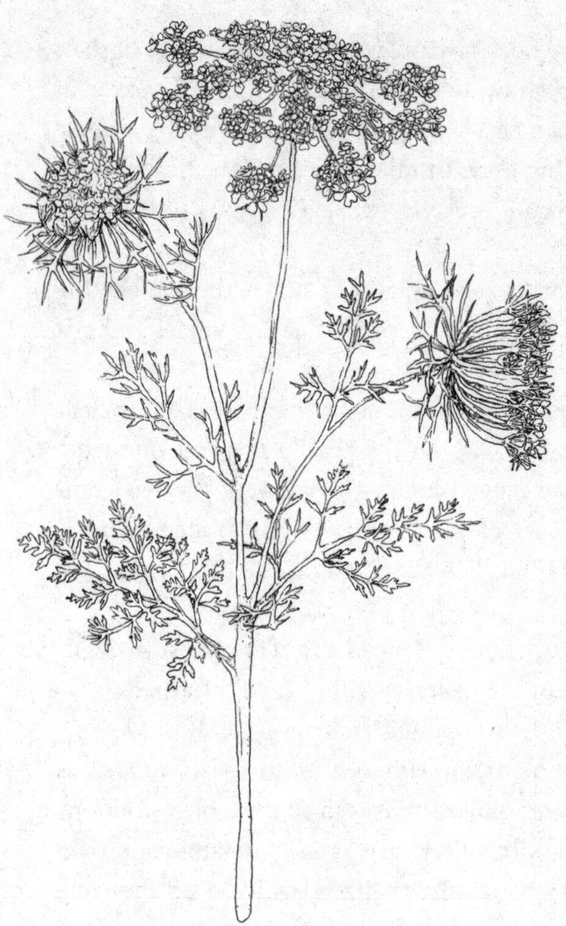

Daucus carota, Queen Anne's Lace, Wild Carrot

When I forage, I tend to find more than *just* what I've set out for, and it is good practice to separate species from one another in your basket. Paper bags are the perfect dividers. Sometimes, even often, you'll come across the unexpected, and you'll want to be able to keep things relatively organized to avoid confusion later.

There are plenty of reasons to avoid plastic—the environment is of course a significant one, but another reason is that mushrooms (with some exceptions) tend to turn slimy when kept in plastic. My recommendation? Ditch the containers and get yourself a fat stack of brown paper lunch sacks. Paper bags are forgiving; they allow for air circulation while absorbing moisture, allowing mushrooms to stay dry and slime-free. When storing mushrooms in the refrigerator, paper will make your mushrooms last significantly longer than plastic.

Maybe mushrooms prefer paper because they are closer with their friends, the trees. It wouldn't surprise me.

Other things that are not mushrooms also keep well in paper bags, like cattail pollen. Cattails are a remarkable, enormously useful plant ally to become acquainted with. The most common species in North America are *Typha latifolia* (the native broadleaf cattail) and *Typha angustifolia* (the invasive Asiatic cattail). Both are equally useful to foragers and wildcrafters. In the Potawatomi language the cattail is known interchangeably by names that translate to "shelter weed" and "we wrap the baby in it," tender monikers that indicate a deep love between this plant and the people who have made use of her many gifts.

A great deal of layman's debate surrounds the legality of cattail harvesting, which can be summed up by the frustrating reality that local laws about foraging can be confusing and impossible to understand, often by design. According to the research I have done, harvesting cattails in North America is generally permissible as long as you are not removing plants from a landowner's property without their knowledge or consent (legal advice that one can assume typically applies to all matters concerning private property), and as long as you are not removing plants from any public land where such actions are prohibited. Some US states may vary slightly in their regulation—sometimes native cattails may be afforded some protection, while invasive cattails are fair game. Other states do not make any distinction between native and invasive.

Part of what makes cattail so very forageable is that every part of it is edible, from the underground rhizomes and laterals to the shoots in spring, the immature green flowers (which can be roasted like corn on the cob and eaten off the stem), even my favorite part, the bright yellow pollen from the male flower.

There are two ways that I know of to collect cattail pollen. Old guidebooks will typically recommend that you grab an empty jar or a bag, hold it up to the plant, tip the yellow flower down into the jar, and shake it vigorously until the pollen falls in. This is, unfortunately, much less efficient than the authors of those guidebooks will ever let on and will not produce nearly enough pollen for the pancakes they're always yammering on about. After an hour or two of work, you may procure a measly half cup or so of pollen, but you will likely have more on yourself than in your container.

The second option (and by far the more practical one) involves snipping the yellow flower off where the stem is separated from the fluff, and putting it into a bag and bringing it home. At this point, the pollen can be laid out to dry on a cookie sheet near a sunny window if necessary, then sifted through a fine-mesh strainer.

Cattails, along with many other aquatic plants, are what are known as "bio-accumulators," meaning they act as living sponges for environmental pollutants, soaking up toxic compounds and heavy metals from the soil and the water and filtering them through their bodies. Because of this generous and self-sacrificial tendency, it is crucial that cattails only be harvested from clean water, far from roads and industry, including agriculture. Otherwise, you could be ingesting these unsafe compounds yourself, and I'm willing to bet that you probably aren't as good at processing them as a cattail is.

I SNAP OFF a dozen canary yellow cattail flower heads into a paper bag, gently shaking a few of the ones I leave behind. Cattails are wind pollinated, and by simulating the wind I can help them to reproduce, which feels like a fair exchange for a few flowers. I laugh and pretend to be a bumblebee resplendent with yellow saddlebags, sending wave after wave of pollen flying through the air.

Wending my way through the reeds, I enter the heart of the bog. Insects buzz around my head, an annoyance, but a worthwhile one. The murky water purrs with life, breathing slowly like a set of mossy, turbid lungs. Islands of sphagnum moss are layered throughout the muck, hosting pitcher plants, leggy orchids, and blueberry bushes resplendent with oblong white flowers.

Bogs and other wetlands, once estimated to cover over 220 million acres in the contiguous United States, have been bisected, with only about 110 million acres remaining. For all our urgency regarding the conservation of trees and forests, little attention is paid to the ecological benefits of wetlands. They are tremendously productive ecosystems, supporting a complex web of biological relationships representing every known taxonomic kingdom. From acidic bogs to alkaline fens, each supports its own delicate and specialized species, many of which cannot be found in any other environment. When a wetland is destroyed, millions suffer.

Until relatively recently, we had deemed wetlands economically worthless, viewing them as unsuitable for both farming and urban infrastructure, and in the case of early settlers, filthy and teeming with disease. But, as we learn time and time again, there is more to the land than meets the uninitiated eye. Indigenous cultures around Turtle Island have viewed wetlands as essential lifeways since time immemorial, vast gardens bursting with food, medicine, materials, and more. In fact, the word *swamp* itself is fairly recent, not even mentioned in written English until the 1600s, when John Smith first used the word in his manuscript *The Generall Historie of Virginia, New England and the Summer Isles.* Such places provided essential shelter for tribes under colonial attack, since they were seen as impossible to inhabit or traverse by colonizers, though in Smith's book he describes them as places of cold, death, and disease, bragging that this was proof of God's righteousness, writing that "as some flourish, others perish." For all Smith's mocking of the so-called primitive people and their inferior ways, he would never understand the wisdom of the swamp.

Places like this are both womb and graveyard, reminiscent of the primordial soup from which our species once emerged. Heavy, anaerobic blankets of decaying peat and loam layered over thousands of years give birth to millions of lives, gurgling and straining to surface above the gloaming depths.

Despite past disregard, the benefits of wetlands cannot be overstated. It feels wrong somehow to laud the economic values of these ecosystems, as if we're missing the point entirely, but even so, they are significant enough to warrant mention. Wetlands dramatically improve water quality, absorbing excess runoff and filtering it before it can enter other waterways. Even a one-acre wetland can store approximately a million gallons of water. Robust coastal wetland ecosystems have been found to protect against erosion and reduce flooding caused by hurricane and tropical storm conditions, and conversely, they conserve water so efficiently that they can act as firebreaks in drought conditions.

Our society tends to judge the value of so-called wilderness by metrics that prioritize extraction economics, and aside from a few crops that can be grown in places like this bog, plenty of people see them as a nuisance more than a benefit. I can understand, on some level. Most Americans don't eat cattails, their blueberries and cranberries come from farms, and their orchids live in greenhouse pots. Wetlands are good habitat for mosquitos and other buzzing insects, they smell a little funny, and you can't build supermarkets on top of them very easily. I'd argue that you shouldn't bother—a healthy bog is all the supermarket I could ask for.

I walk off the end of the boardwalk into the forest on the other side of the bog wearing smears of yellow dust on my overalls, feeling a sudden change in my knees as the bouncy, supple bridge gives way to solid earth beneath my feet. I rustle through fluttery lamb's-quarters, *Chenopodium album*, a quinoa relative that tastes like sour spinach. The leaves are covered in a soft, white powder, which brushes onto my clothes and mingles with the yellow pollen. The loud smell of bog persists, but other scents march forward in the olfactory symphony—flora, a faint fruitiness, and something fungal.

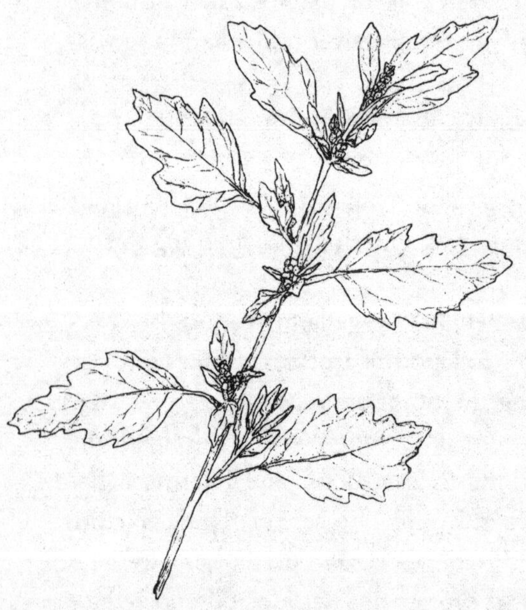

***Chenopodium album*, Lamb's-quarters, Goosefoot**

This forest is utterly rich with mushrooms. Russula have erupted in dozens of shades of pink, red, and purple, a genus notorious for massive species diversity and absurdly nightmarish identification qualifications.

The word *russula* comes from a Latin word meaning "red," although russula are by no means always red or some shade thereof. Their typical characteristics are a brightly colored pileus with a depressed center (like a belly button) and very brittle gills that are distinctly attached to a stout stipe. They do not "bleed" milky latex like their close relatives in the *Lactarius* and *Lactifluus* genera, nor do they possess a "ring" (called an "annulus") beneath their gills.

They are common, and there are hundreds of *Russula* fungi, some estimates reaching as high as 750 distinct species, each one more difficult than the next to speciate by visual characteristics alone. As Michael Kuo says, "Russula identification is a joke," at best an endeavor that can only result in drowning under the weight of pointless and variable criteria that cannot generally be quantified, as individual specimens may display wildly different characteristics despite being the same species. Perhaps even more frustrating, completely different species may appear identical to even the most highly trained eye.

Many mycophagists do not bother figuring out which *Russula* they have found, thanks to their ever-changing and often contradictory criteria, and instead perceive individual edibility or inedibility by performing what is colloquially known as a "nibble-spit test."

A "nibble-spit test" is a simple field exercise so named for its process: tasting a small piece of the mushroom, letting it sit on your tongue for a moment, and then spitting it out. For *Russulaceae*, including genera *Russula*, *Lactarius*, and *Lactifluus*, pleasant or sweet flavors confirm an edible species, but even spicy or bitter species are quite likely to be edible, save a few select species (such as *Russula* subsp. *nigricantes*, the blackening brittlegill). Spicy *Lactarius*, for example, will mellow out when cooked, and extreme or unpleasant bitterness is often rescued by fermentation.

Mushrooming is a sensory activity, extremely so. No part of you can be disengaged, not for a moment. Inattentive mushroom hunters make mistakes, even those with strong knowledge. They get lazy, sloppy, they miss steps, sometimes even stop thinking altogether.

If you want to shut your brain off for a while, go pick raspberries.

Don't go picking mushrooms if you can't be present with them.

I leave these mushrooms alone. No offense to *Russulaceae*, I just don't particularly enjoy eating most of them. I'll eat them if necessary, or if they are prepared in certain ways, but this time, I want to save some room in my basket for other treasures, and the air is getting sweeter as I press farther up and farther in.

I know what's waiting for me.

Whenever I am in a position to be giving a talk about mushrooms, during the Q&A session I will inevitably be asked one enduring, completely predictable question: "What is your favorite mushroom?"

If you are asked this question enough times, even if you feel, as I do, that it is increasingly impossible to answer, you try to recognize the spirit of the question and give a decent response. The best answer, which I will forever wish I had thought of myself, was originally given by the mycologist Alan Rockefeller, who simply replied, "Whatever mushroom is in front of me at the time." However, due to my reluctance to plagiarize, I typically respond somewhat sheepishly with "chanterelles," which, to be fair, isn't entirely untrue.

Why chanterelles? Well, a few reasons. For starters, they are one of the most ambidextrous mushrooms I can think of. What other mushroom has the liminality, the sheer *range* to be the main character in ice cream *and* in pasta? Plus, chanterelles are pure joy to look for. Bright yellow, scattered across the forest floor like fallen doubloons from a pirate's overflowing treasure chest, with a sweet, fruity fragrance of peaches and apricots, bathed in a sort of floral perfume that I've never been able to put my finger on. They are firm but not unyielding, assertive but not overbearing in flavor, a mushroom that can only be described as "agreeable." Ranging from the characteristic golden yellow to bright red, with some varieties toward the West Coast sporting white, blue, and even "rainbow," each is remarkable in its beauty. Most species (shy of the "smooth" chanterelle, *Cantharellus lateritius*) present with distinct decurrent ridges along the stipe, typically forking as they approach the edge of the cap, and if you look closely at a few, you will see that each pattern of ridges looks just a little different from the last, giving the impression of a sort of fungal fingerprint. The best of them all, in my opinion, *Cantharellus phas-*

matis, are pale, shaped like odd, chunky baby feet, looking almost overinflated.

Of the entire genus (which contains a multitude of species, both described and undescribed), golden chanterelles are the most recognizable, generous, and among the most abundant.

If you look in the right place at the right time during a summer with enough rain, you could fill a backyard swimming pool with an ocean of perfect golden chanterelles and the forest might very well still look untouched.

It seems reasonable, then, that if you follow the rainbow, treasure may very well appear at your feet, as it seems, it has appeared at mine.

All around me, seas of golden mushrooms.

JUNE 2022
Kalamazoo, Michigan

No matter the environment, be it wetland, forest, city, desert, or tundra, foraging is almost always possible. I have received countless emails and messages from those who consume my online content, all saying something akin to "it's all well and good for you, you live in the woods! I live in a city/desert/outer space where there's *nothing* to forage!" It is usually at this point that I am in the happy position of informing these unhappy campers that actually, some of the best foraging I do is not deep in some lush forest, but in concrete jungles (and contrary to their assumptions, I *do* happen to live in a city). You see, the key to finding more food isn't getting farther away from people. More often than not, moving *closer* to human epicenters will net you more substantial, varied, consistent, and predictable foods.

It makes sense. People tend to concentrate in specific areas, and those areas are usually chosen for a reason. Urban centers are frequently built around natural features like rivers and lakes, contain green spaces like walking or bike trails, parks, and cemeteries, which tend to harbor biodiversity, and perhaps most importantly, attract people from a large variety of backgrounds, cultures, places of origin, and traditions. When people move to new places, they tend to bring their most important plants with them. Food plants in landscaping usually have some correlation to the presence of people who use them. The long scar of the Trail of Tears can be marked by plants that Native people carried with them as they were forced out of their homelands. If you walk a portion of the trail today, you might find some of them. Hell, sometimes edible plants even end up in the landscaping with no intention of being food, but just because they're pretty. When I visit New York City, I notice the gingko trees loaded down with fruit in Chinatown, grapes winding up the chain-link fences around basketball courts in Harlem, burdock flourishing in Central Park, even apples, pears, nannyberry, and Cornelian cherries all in the middle of the city. Oak trees still rain down acorns even if they only land on the sidewalk. Oyster mushrooms are happy to grow from telephone poles, and agarics still thrive in mulch, even though they were never planted there.

Plants love us. Fungi love us.

We evolved alongside one another, and we flourish when we have one another.

One of the most hotly anticipated fruits of the forager's year is one that is hiding, quite literally, in plain sight in many of our urban environments. While they do grow wild, this native fruit is frequently planted as a landscape tree, beloved for its beautiful flowering stage, red-purple fruits, and yellow-orange-red leaves in autumn. It goes by many names, including saskatoon berry, Juneberry, *Amelanchier*, and serviceberry, and while it can be found in forests, its edible fruits are noticeably improved in both quality and quantity by their presence in urban landscapes.

"What are you doing?" one concerned person after another asks me as I cheerfully gather pounds of sweet red berries from the trees outside of Trader Joe's. "Picking berries," I respond, "happy to share."

"No thanks," they say, as they hustle from me with their paper bags full of paid-for groceries while I fill a basket for nothing but the price of my time.

I shrug. Something about horses and water comes to mind, but my mouth is too full of fruit to bother with coaxing anyone else to join me.

Landscape fruit is like that. People are suspicious of anything "ornamental," assuming that the word is synonymous with "inedible." I suppose we're used to not being able to trust anything free in this world, especially when its intended purpose is beauty, not sustenance. Fortunately, plants often manage to be both at once.

The birds and I continue to pick from the tree together. They go high, munching on the fruits I cannot reach, and I collect lower, listening to little berry thumps with significant pleasure as the basket continues to fill up.

I feel a tap on my arm and look down to see a little girl with giant brown eyes and chunky red and purple plastic beads in her hair. "*Excusemewhatareyoudoing*," she says, all in one breath without a question mark at the end. I look behind her briefly and see a woman, likely her mother, standing a few feet away, giving the sort of prompting eyebrow look to the little girl that says "You wanna know, you gotta go ask yourself."

I respond carefully. "Hi! I'm picking berries. The grocery store planted them just to be a pretty tree, but a lot of people around the world eat them for food. They're called Juneberries. See how they look kind of like a blueberry, but they're a different color?" I bend down and show her the crown at the top of the berry. "See those little flaps on top, shaped like a star? There are five of them, you can count them."

We count in unison, "One, two, three four five."

"Blueberries have them too."

"But they're red. And purple. Not blue," she says with a serious look on her face.

"You're right. The same color as your beads. Hey! Maybe you're a Juneberry tree too!" I joke, and her face scrunches up in giggles.

I acknowledge the woman behind her with a wave. She comes over and looks with interest in my basket. "So they're edible?"

"Edible and fantastic. I pick loads every year all over town." I reply.

"Really!" she exclaims, a brief look of skepticism crossing her face.

"Really. Would you like to try?" I ask, casually popping a few in my mouth as an endorsement for their safety and holding my basket out to her. (I've found that people are more willing to try from the basket than the tree.)

Hesitant but too curious to say no, she reaches in and selects a berry, tasting it. Her eyes widen, and her expression relaxes. "They're delicious!" She reaches up in the tree and picks a few, then hands them to her daughter. "Do you want to try?"

The little girl nods and accepts the berries. Her giant, gap-toothed smile says it all. They're both sold.

We pick berries together for the next stretch of time. They fill two paper lunch sacks from my car, or, at the very least, Mom fills hers. The little girl prefers the "built-in bucket" as I like to call it, staining her tongue and teeth purple.

With the three of us there chatting, eating, collecting, and laughing, we're not getting nearly as many sideways glances from passersby as I was on my own. If anything, people seem more oblivious to us, and those who do notice us seem more curious than suspicious, like they're afraid they might be missing out on something good. They say nothing is better for business than a line out the front door, and the same seems to hold true for urban foraging.

Within twenty minutes, a small but enthusiastic crowd forms. Some people only stop by for a taste before getting on their way, a few others stick around and fill whatever vessels they can find—Starbucks cups, reusable water bottles, empty hands. In our own way, we are temporarily free of the plastic clamshells with price stickers on the other side of the sliding glass doors, not ruled by money or jobs to earn our berries.

Foraging is more than a hobby or an idle pastime no longer necessary for survival. It is in our blood, an evolutionary compulsion that the crushing conveniences of the modern age have not yet bred out of us. There is a part of us, a deep and curious part, that longs to let our senses run wild, to pass beyond the pressures of clock time and perform useful labor with immediate and satisfying results. We all long to grow and explore, to be curious, to pick berries in the sun in good company.

Once upon a time, before modern agriculture, before occupations, when the world was still young and innocent, we were all berry pickers, sticky hand to purple mouth.

JUNE 2020
Indianapolis, Indiana

My family has a tradition when we travel together. Whenever we cross state lines, you're bound to hear us bid goodbye to the state we're leaving and hello to the state we're entering. Living in the tri-state area, cries of "Hello, New York" and "Hello, New Jersey" could often be heard from the back of the car, usually coupled with "Goodbye, Pennsylvania."

The RV is clean and empty, and the truck is packed. All the months we've spent together struggling through what felt like endless problems with exploding water lines, broken electrical boxes, an elusive septic tank, persistent leaks, and snowdrifts that threatened to cave the roof in suddenly feel like a barely remembered dream.

I don't want to forget. This place is my triumph, my proof to myself that I have always been strong and capable enough to live my life on my own terms, to build back from nothing. To not just survive, but to thrive. My eternal *fuck you* to that mocking voice in my head, the ghost that haunts me with frightening predictions about my future.

I cut my teeth in this place.

That has to mean *something.*

My new apartment is waiting for me in Michigan, a tiny but affordable place in a secure building with a working shower, toilet, AC unit, heater, and sink. Luxury beyond my wildest dreams, as far as I'm concerned.

With the pandemic in full swing, everything feels uncertain. What will it be like to start my master's degree online, with the university just six minutes down the road, but totally empty? I am as alone as I have ever been, barely a single contact in Kalamazoo apart from the professors, with no prospects of meeting anyone or making any friends. It feels like a blank slate in front of me, and behind me, a decade of history, humans, and connections and places seared in my chest.

Overnight, it will all be four hours behind me.

No looking back.

Typically when I'm this anxious, I go to work. But I've already turned in everything, received my last paycheck, handed all my duties over to extremely capable new hands.

It isn't my job anymore.

It also isn't my house anymore. Someone will come and take it away with a pickup truck tomorrow, and grass will grow back.

Pretty soon, it will look like I was never here at all.

I get in my car and drive until I reach Indianapolis. I don't know what else to do. Everything has been shut down, and the only cars on the road belong to people who have no choice but to be there.

Lost in thought, I drive as if on autopilot until it feels right to stop.

When I arrive, Broad Ripple Park is a ghost town. The few people out walking, jogging, and playing with their kids keep to themselves, masked up and distanced. The playground equipment is abandoned, and children who try in vain to approach it are shooed away by anxious parents brandishing Lysol wipes.

I have no basket with me. Practically everything I own is packed up in the truck already, and for today, I have no home to return to.

Both of my hands feel uncomfortably free as I walk the path, listening to sounds I never remembered hearing in this park before. The scratching sound of a squirrel's toenails as it runs up an old elm, the gentle little *thumps* made by birds dive-bombing insects in the grass; here, absolutely everything is amplified by the sudden lack of human activity.

Memories flood me. This was where I found my first giant puffball, just over to the right, by the fence line. I thought it was a volleyball before I realized it was a mushroom. Over there, just by the spruce tree, a hidden stump under the brush. I found my first chicken of the woods there, twenty pounds of it. I carried it three miles back to the house in my sweatshirt. Over through the trees, along the path—that's where the wild ginger, *Asarum canadense*, grows. Its dense thicket of rhizomes knits a warm blanket along the forest floor, sweet and spicy, resplendent with fashionably glossy heart-shaped leaves. The first thing I ever made with wild ginger was a simple syrup that I added to cocktails.

I was so proud that something I foraged and then made into something else actually tasted good.

Prunus americana: wild plums. A thicket of stubby trees, each one producing dozens of pounds of small purple, red, and yellow fruits for a few fleeting weeks in the summer. Blink and you'll miss the whole season, but if you're impatient, the green fruits can be infused in alcohol and used to make a light, refreshing liqueur.

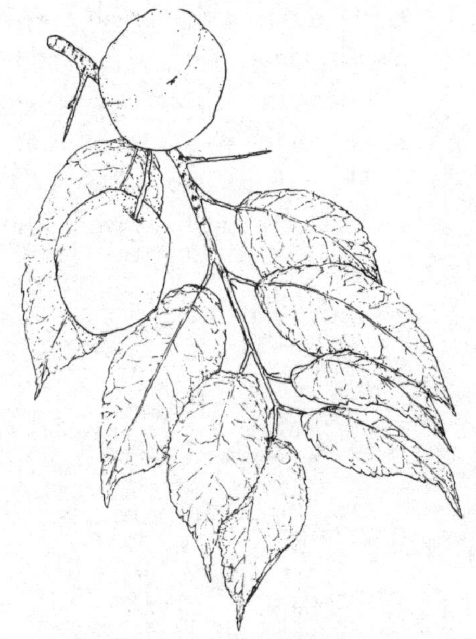

***Prunus americana*, American Plum, Wild Plum**

The logs—they grow resinous polypore—*Ischnoderma resinosum*. *Ischnoderma* comes from two Greek words: *ischnos*, meaning "wrinkled or withered," and *derma*, meaning "skin." *Resinosum* means, as you might guess, "resinous," referring to this polypore's distinctive reddish-brown guttation (a sort of mushroom "sweat" that can be found seeping out of the pore surface in thick droplets on young or quick-growing specimens). They were called "steak of the woods" by someone on a foraging forum I frequented. I found my first ones on that tree, right there. I sliced the tender mushrooms cleanly from a rotting tree trunk in the rain and cooked them up with onion and garlic at home.

My old home. The one before this, with the green door and the stove with only one working burner.

The tree is barren now, not a trace of fungi upon it. It is spring, not winter. Nearly summer, as indicated by the spreading tree canopy and the sun's heat on my shoulders. The resinous polypore are all sleeping.

That liminal space between spring and summer usually feels like my first day on a new job; the jitters and anxiety have been there, but the actual work has been theoretical until now, when suddenly, everything starts moving faster. The lengthening days have a quickening effect on the foraging season, and everything is exploding at once. So much work to do, and in so little time.

I don't think I can be here right now. I need to move on. There is too much to say goodbye to, and so little time.

The walk back to the car feels as though I am watching someone else take over my body while I float above it, watching myself like an avatar.

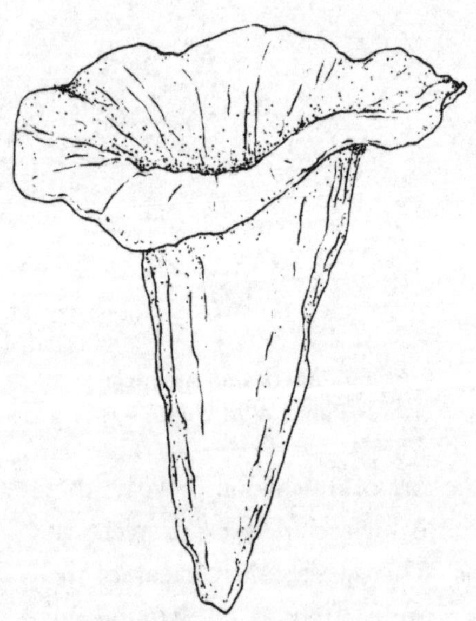

Craterellus sp., Black Trumpets, Black Chanterelles

Next up is Holliday Park. A tough one for me.

I found my first edible mushrooms just there, to the left, down by the bottom of the steps. To the right, down the path just a bit, I found my first pawpaw, *Asimina triloba*, back in the fall of 2013. I was impatient and didn't wait for it to fall on its own, didn't know it wouldn't ripen after I took it from the tree. But I learned. I grew. I found my way.

Over there, I collected black trumpets, *Craterellus fallax*, on the ridge in the summer of 2018, a shock of flimsy, gray-black mushrooms with deep hollows where one might expect to see a cap, like slender vases. I collected this unexpected treasure in such abundance that I ended up back at my car in my sports bra, my T-shirt serving as a makeshift bag.

Tiny thorns snag on my clothes as I push through the edge of the forest. Black raspberry and blackberry, *Rubus occidentalis* and *Rubus allegheniensis*. Sweet, seedy, sun-warmed. I can imagine the taste, hold it in my mind.

Up by the parking lot, the mislabeled hickory tree. Shagbark, not bitternut. I collected nuts there in the fall anytime they masted, bringing them back to my apartment in trash bags and processing them on the back porch with a hammer and a nutpick.

My memory of my time here is full of contradictions. I had friends, mentors, little pockets of peace and joy in parks and secret hollows around the city. Homes I grew to love, places where I could find a little wildness. Still, I felt lonely, trapped. I didn't belong here forever—I was certain that I would leave after college and explore the rest of the world, go to grad school, make something bigger than this place could contain. When that didn't happen, I spun my wheels, made myself small. Settled. Rotted.

And yet, even though I know I need to move on, to chase what my heart has always known, that Indiana was never meant to be my home forever, I still cannot help but hold a space for those precious moments of wonder that I experienced here. My first wild mushrooms. Sweet, custardy pawpaw mash and dark red berry juices dripping down my chin. Stuffing autumn nuts in my pockets until the fabric threatened to rip. The dilapidated places I have transformed into homes. The world I learned to notice, to hear, to recognize within the confines of this city.

Despite all the bitterness I've held in my heart for this place, it still seems to have left a mark on me.

The wind is shifting, and the season with it. A fresh new breeze from the north, sweet and clean. *Go on,* it seems to whisper, *go find the place I come from.*

The song of the forest rises with the wind in a chorus, sending a symphony of forest sounds and smells cascading around my body. *It's time to go. It's time to grow.*

Goodbye, shellbark.

Goodbye, pawpaws.

Goodbye, honey mushrooms.

Goodbye, wild ginger.

I touch the handle on the car door and look behind me.

Thank you. For everything. I whisper.

Goodbye, Indiana.

Hello, Michigan.

GATHERING EXERCISE

"SYMPHONY"

---

* Take a notebook out to the forest with you, choose a spot, and write down everything you can smell from a single position. Try to collect at least five scents (this may take some time and focus, especially if you are unaccustomed to noticing smells).
* Think of a musical instrument that you can associate with each smell. Perhaps the resinous scent of conifer reminds you of a cello, or the smell of lichen makes you think of guitars. If this is too difficult, imagine what instrument the origin of each of these smells would choose to play if they could.
* Imagine the music that the smells would play together in your head, and enjoy it for a while. Consider writing down some of your thoughts about what you have "heard."

Umeshu

Makes 1 quart

Once per year, I collect unripe green plums and use them to make a Japanese beverage called *umeshu,* known in English as "plum wine." Technically, the *ume* plums used in Japan aren't true plums at all, they're more like apricots, but this recipe will work with any unripe stone fruit, from peaches to nectarines, apricots, and, yes, even actual plums. Umeshu also isn't technically a wine at all; it's more of an infused liqueur (when translated, the word *umeshu* simply means "ume alcohol"). It can be enjoyed after six months, but it's much better if you wait at least a year for the more delicate, subtle flavors to emerge.

To make this recipe, you'll need to procure some fairly neutral high-proof spirits. The higher the proof, the faster it will extract your flavors, but you may find that something with too high a proof is too undrinkable on the back end. Sake is a popular choice, and vodka another. You'll also need to obtain some rock sugar, which you can find at most Asian grocery stores. The size of the sugar crystals in rock sugar prevents them from dissolving too quickly, which allows for a much more fragrant and complex extraction of the plum flavors.

1 quart-size mason jar
1 cup unripe green plums, washed and dried
1 cup rock sugar
2 to 3 cups neutral high-proof spirits

In your mason jar, add a single layer of plums to cover the bottom. On top, add one layer of rock sugar, about the same volume as the plums. Repeat until the jar is full, but not packed, with at least two inches of headroom. End with a layer of sugar.

Pour your alcohol into the jar until 1 inch of headroom is left, then seal with a lid and a ring. Store in a dark, cool place out of the way for at least 6 months.

(continued)

After six months, taste your umeshu. The rock sugar should be partially dissolved, with shimmery trails of dissolved sugar visible in the jar, and the plums may appear to be smaller and slightly wrinkled. Seal it up and return it to a dark, cool place for another 6 months.

After 1 full year has passed, the sugar should be fully dissolved and the plums should appear much smaller and very wrinkled, almost prune-like. They can be left in the jar with the liqueur indefinitely, or they can be removed and eaten on their own. Serve your umeshu as a mixer with sparkling water and ice, or top with champagne or prosecco.

Cattail Pollen Truffles

Makes about 24 truffles

Pollen in general is an unfamiliar ingredient in most kitchens, but it can be a lovely way to add a touch of surprise to an otherwise ordinary recipe. Cattail pollen has an earthy, dusky flavor that may remind you a little bit of sweet corn, with a luxurious, silky texture. I like using cattail pollen because of its complex flavor and heat-stable, canary-yellow coloring, and I find it transforms a basic chocolate truffle into something spectacular and unexpected. Other plants with edible pollen, such as spruce or pine, are less colorful and may be a bit less allergy-friendly due to the smaller size of the pollen grains. Cattail pollen has comparatively large grains and is unlikely to cause respiratory issues in most people, but try not to breathe it in as you gather it to be on the safe side.

Cattail flowers have both pollen-producing and pollen-receiving parts. The pollen-producing parts are situated at the top of the plant, with pollen receivers directly beneath them. Since cattails are wind-pollinated, even a light breeze will send pollen cascading down onto the pollen receivers of nearby flowers. To harvest cattail pollen, the most efficient way is to collect the top part of the maturing flowerhead by snapping it off the plant and into a plastic or paper bag. Collect mindfully (you don't need much, maybe 15 to 20 pollen heads for a quarter to half a cup of pollen) and try to be messy about it! "Wind-pollinated" can also mean "YOU-pollinated!"

When you've collected your flowers, you can process them by smacking your bag around on nearby trees or rocks to dislodge the pollen. When you get home, use decreasing sizes of tight-woven mesh sieves to filter out fluff from pollen until you are left with a soft, fine yellow powder. Store it in a plastic bag or a mason jar in the freezer until you are ready to use it.

8 ounces dark chocolate (I use at least 70%, but this can be adjusted to your preference)

½ to ⅔ cup heavy cream

1 tablespoon unsalted butter

⅛ to ¼ cup cattail pollen

Chop the chocolate into small pieces with a heavy knife, then place in a heavy heat-safe bowl. Warm the heavy cream in a small pot until it is between 150 and 160°F, or until steam begins to rise, bubbles appear at the edges, and the cream is slightly uncomfortably hot to the touch. Immediately pour the hot cream over the chocolate, add the butter, and fold together until silky smooth.

Place the mixture in the refrigerator for 30 to 60 minutes, until firm yet pliable enough to scoop easily with a spoon. Remove from the refrigerator, shape into balls of relatively equal size (mine are about 2 tablespoons each, but you can make them larger or smaller). I use clean, ungloved hands for this step, as I find that the warmth from my hands helps to create a smoother exterior texture, but you can also use a dough scoop or wear gloves if you don't prefer the more "rustic" touch. Decorate with pollen, either by rolling to coat or by sprinkling on top. I occasionally pop these in my lunch bag for a little pick-me-up in the middle of a long day, and they also make fantastic homemade gifts for the adventurous foodie in your life. The truffles can be stored in the refrigerator for up to 3 weeks (if they last that long).

Chanterelle Peach Pie with Basil

Serves 4 to 6

Mushroom *pie*?! Believe it or not, chanterelles are delicious in both sweet and savory applications. Though adding mushrooms and basil to peach cobbler might seem like a tough sell, you'll be shocked by how good it is. The chanterelles, in addition to their lighter fruity and floral notes, lend a more complex texture and an unexpected peppery zing that steps in parallel with the robust, juicy sweetness and tartness from the peaches, and the basil bridges the gap between the two seemingly disparate ingredients like a charm.

If you have a family biscuit recipe, feel free to substitute with your own—I don't want your granny on my case!

For the filling:

⅓ cup brown sugar
1 teaspoon salt
Juice of 1 lemon
1 splash vanilla extract
2 tablespoons cornstarch or arrowroot powder
4 tablespoons (½ stick) unsalted butter, melted
½ pound chanterelles, chopped or torn into 1-inch pieces
4 to 5 large ripe peaches, pitted and sliced or cut into 1-inch pieces
Small handful of fresh basil leaves (do not use dried), torn

For the biscuits:

2 cups all-purpose flour
⅓ cup granulated sugar
2 teaspoons baking powder
¼ teaspoon baking soda
Heavy pinch of kosher salt, about ½ teaspoon
½ cup (1 stick) unsalted butter, frozen, plus more for greasing
½ cup very cold buttermilk, plus more if needed

Preheat the oven to 350°F. Grease a 10-inch baking pan with butter.

First, make the filling. Combine the brown sugar, salt, lemon juice, vanilla, cornstarch, and butter in a pot and heat over low heat until the mixture starts to bubble. Add the chanterelles and peaches to the pot and cook for at least five minutes or until the liquid is thick, syrupy, and fragrant. Set aside to cool.

Prepare the biscuit dough: Sift the flour, granulated sugar, baking powder, baking soda, and salt into a large bowl. Grate the frozen butter in, working quickly to avoid letting the butter warm up too much. Pour the buttermilk in a little at a time and stir with your hands until you have a wet, slightly sticky dough with dots and streaks of visible butter throughout but no obvious pockets of dry ingredients. Be careful not to let your butter melt from the heat of your hands. If your dough is too dry, add another splash of buttermilk, but don't overwork it!

Add the filling to the prepared baking pan and sprinkle the basil on top. Take chunks of biscuit dough and place them on top of your filling, forming a relatively even layer of biscuit dough that mostly covers the filling. Don't overthink it.

Bake for 40 minutes, or until the top has turned golden brown. Cool just until no longer piping hot, then serve with a scoop of ice cream or lots of homemade whipped cream.

Lacto-Fermented Black Trumpets and Cultured Butter

Fermenting is an excellent method for processing many wild mushrooms. It preserves them for later, brings out their best qualities, and may even make them more easily digestible, since fermentation breaks down some of their more complex structures. Black trumpets have surprising notes of chocolate and smoke that can be enhanced by fermentation, but you can ferment lots of other mushrooms (including chanterelles). Be warned, however—some mushrooms, such as morels, *require* cooking to be considered safe for human consumption, as they contain volatile compounds that are destroyed by heat but not necessarily through the fermentation process.

This recipe makes two wonderful products in one (technically three, if you count the buttermilk). We start by making lacto-fermented black trumpet mushrooms, which can be eaten on anything from pasta to pizza to charcuterie. But as if that weren't enough, we then take the liquid by-product of the black trumpet fermentation and use it to make one of my very favorite things: cultured butter, infused with all the complex and beautiful flavors of our mushrooms.

To ferment mushrooms, we'll use the same process we used earlier for our garlic mustard (and we'll be doing it again in just a few pages with our Juneberries), so you don't have to change very much, but I'll reiterate below. When working with mushrooms, instead of using a 2% ratio of salt by weight, I bump it up to 3% to account for the higher pH. This recipe assumes that you have all the things referenced in the recipe for lacto-fermented garlic mustard (page 78).

Part 1: Lacto-Fermented Black Trumpets

Black trumpets
Non-iodized salt

Part 2: Black Trumpet Cultured Butter

A little less than 1 quart pasteurized heavy cream
2 tablespoons reserved fermentation liquid (a by-product of part 1)

Part 1: Lacto-Fermented Black Trumpets

Weigh your mason jar, hit the "tare" button, then fill your jar with mushrooms (leaving a few inches of headroom for your weight). Note the weight, then tare again. Add 3% of the weight of the mushrooms to the jar in salt. Shake vigorously until the salt coats the mushrooms. Add your fermentation weight, then seal the lid. The lid should be closed, but not so tightly that it is difficult to open, since the gases in the jar will make it harder to unscrew once fermentation has begun. Leave the jar in a cupboard or some other warm, dark place, and burp the jar once per day for 3 to 5 days.

When the jar of mushrooms is fragrant and the mushrooms appear to be floating in liquid, they are ready to be eaten. Store in the refrigerator to slow down any further fermentation.

Part 2: Black Trumpet Cultured Butter

Pour the heavy cream into a clean quart-sized mason jar, leaving 2 inches of headroom from the top. Reserve 2 tablespoons of liquid from the jar of lacto-fermented black trumpets, then add to the jar with the cream. Seal finger-tight, then rock the jar upside down and side-to-side to fully incorporate the fermentation liquid into the cream, avoiding any vigorous shaking. Set aside in a warm place, such as on top of your refrigerator, for 24 hours.

After 24 hours, the cream should be significantly thickened, about the texture of whipped cream. If you have a stand mixer, dump the cream out into the bowl and, using your whisk attachment, spin on high until the buttermilk and solids separate, about 8 to 10 minutes. If you do not have a stand mixer, you can either use a hand mixer and a bowl or you can simply shake the mason jar (keep it firmly closed) until the butter visibly "breaks" into distinctive buttermilk and butter solids.

Prepare a large mixing bowl with high sides. Strain out the buttermilk and discard or set aside for later use. Dump the butter solids into the bowl. "Wash" the butter by rinsing it in cold water until it runs clear, then strain the water out. Using a spatula, squish the butter up against

(continued)

the sides of the bowl, squeezing out any remaining liquid. Discard all remaining liquid and store either in a mason jar or in tightly rolled parchment in the refrigerator. Enjoy on flatbread, toast, sandwiches, biscuits, shortbread, pasta dishes, or any other recipe where butter shines!

Thai Chanterelle Red Curry

Serves 4 to 6

While you usually see chanterelles used in more carb-heavy European contexts (pizza, pasta, toast, etc.), I *love* Thai food, so I'm always finding ways to incorporate these lovely mushrooms in the curries that frequent my table. I picked up this base curry recipe a few years back on Reddit of all places while trying desperately to find good step-by-step instructions for layering flavors in curries (and an explanation for why my curries always came out bland!), then added a few twists of my own. Thanks for the tips, @cheftell1921!

If you prefer curries of the "face-melting, rear end destroying" variety, try toasting some Thai red chiles with your coconut cream at the beginning.

1 cup full-fat coconut cream

Dried Thai red chiles to taste (optional)

1 (4-ounce) can Thai red curry paste (I like Maesri)

1 (15-ounce) can full-fat coconut milk

1 knob galangal, grated, about 3 tablespoons (use ginger if you can't find this, but double the amount)

2 tablespoons lemongrass paste

3 garlic cloves, minced

3 to 5 makrut lime leaves (typically found in the frozen section of Asian markets)

½ pound chanterelles (leave intact if small; tear into strips if large)

1 to 2 cups chopped seasonal vegetables (zucchini, bell peppers, cherry tomatoes, etc.), sliced into even chunks about the same size as the chanterelles

1 tablespoon fish sauce (adjust to your taste)

½ teaspoon soy sauce

Juice of 1 large lime

1 small bunch Thai basil

3 to 4 fresh Queen Anne's lace flowers, clipped off the stem

2 cups jasmine rice, cooked

Reduce the coconut cream (along with the dried Thai chiles, if using) in a saucepan over medium heat until the cream breaks (you will see an oily layer begin to form on the top and the cream will begin to look chunky), then add the entire can of curry paste. Whisk it together. This will re-emulsify and look smooth again, but only for a moment! Continue to cook the mixture down until the layers separate again. You will think you have gone too far, that it looks bad. Trust the process—you are concentrating, deepening, and enlivening the flavors when you do this, and you will fix the way it looks in just a few minutes.

Once everything looks so awful and chunky that you don't think you can stand it, but before it starts to burn, add the coconut milk and whisk until it is smooth again. Add the galangal, lemongrass, garlic, and lime leaves. Bring to a simmer and simmer for about 15 minutes until thick and creamy, then add the chanterelles and vegetables. Continue simmering for an additional 5 to 10 minutes, until the vegetables are softened.

Remove from the heat and add the fish sauce, soy sauce, lime juice, and Thai basil. Garnish with Queen Anne's lace flowers and serve over warm jasmine rice.

Lacto-Fermented Juneberry Barbecue Sauce

Makes 1 to 2 quarts

One of my favorite ways to preserve berries is to lacto-ferment them. We did a bit of lacto-fermenting earlier on in the year when we made our garlic mustard pesto (page 78), and just a few pages ago with our black trumpets (page 146), but when we apply the same technique to fresh berries, they get less "stinky" and more fragrant and perfumy. The sweetness of the berries is enhanced by the salty tang of the fermentation process, and it is a creative way to prevent large harvests of berries from rotting in the fridge. Berries that have been lacto-fermented can be eaten just like regular fresh berries—use them to top yogurt or ice cream, in pies and cakes, or just eat them by the spoonful—but in this recipe, we'll be using Juneberries (that gorgeous purple-y fruit I picked outside of Trader Joe's—see page 132) and making one of America's most ubiquitous condiments: barbecue sauce.

To lacto-ferment the berries:

A digital gram scale

A mason jar

Juneberries (if you don't have access to Juneberries, substitute with blueberries)

Non-iodized salt

A fermentation weight or a zip-top bag full of water

Weigh your mason jar, hit the "tare" button, then fill your jar with berries (leaving a few inches of headroom for your weight). Note the weight, then tare again. Add 2% of the weight of the berries to the jar in salt. Shake vigorously until the salt coats the berries. Add your fermentation weight, then seal the lid. The lid should be closed, but not so tightly that it is difficult to open, since the gases in the jar will make it harder to unscrew once fermentation has begun. Leave the jar in a cupboard or some other warm, dark place, and burp the jar once per day for 3 to 5 days.

To make the barbecue sauce:

1 cup lacto-fermented Juneberries
1½ cups ketchup
½ cup apple cider vinegar
½ cup water
¼ cup packed brown sugar
¼ cup soy sauce
1 teaspoon garlic powder
1 teaspoon onion powder
1 teaspoon ground cumin
1 teaspoon smoked paprika
¼ teaspoon liquid smoke

Combine all the ingredients in a blender and blend until smooth. Transfer to a saucepan and cook down over medium-low heat for 15 to 20 minutes, stirring occasionally, until reduced by half. If you plan on canning your barbecue sauce, fill your jars when the sauce reaches the consistency of thick applesauce and leave ½ inch of headspace. Because of the high acidity of barbecue sauce, it can be safely water bath canned for 45 minutes in pint jars. If you do not plan on canning your sauce, cook it down a little further, until it reaches a thickness between that of tomato sauce and tomato paste, at which point it can be stored in jars in the refrigerator for up to 6 months.

Elderflower Honey

Makes 1 pint

I always know I'm in for a treat when I visit a new coffee shop and they have elderflower syrup available, especially if it's made in-house, but did you know that elderflower also infuses beautifully into honey? The honey acts as a natural preservative, continuing to infuse with flavor for months on end without spoiling. I like adding elderflower honey to my coffee, and it's one of those treats I never seem to have quite enough of on hand.

When you're gathering your elderflowers, try to get out there in the morning just as the dew has evaporated but before the flowers have been visited by too many insect friends, as most of the flavor is concentrated in the pollen. Sometimes you can shake or lightly smack the flowers right into your gathering container without having to remove them from the plant, but if necessary, simply snip a few flower heads off with scissors or pruners and leave the rest behind to continue maturing. You can return in a couple months to collect ripe elderberries, which are a great treat for birds and people alike.

1 pint-sized mason jar
1 cup elderflowers (unwashed, plucked away from the stems)
2 cups honey

Fill a pint-sized mason jar with the elderflowers and pour the honey on top. Every day for three weeks, flip the jar over to the other end (one day it will be upside-down, the next day it will be right-side-up). After three weeks, the elderflower scent will have fully infused into the honey, and you can reduce your jar-flipping to once per week or less. After six weeks, using a strainer, separate the honey from the elderflowers. Discard the flowers, and as long as the honey is kept sealed between uses, the elderflower scent and flavor will remain.

Blueberries, Barefoot

If he thinks all the fruit that
grows wild is for him,
He'll find he's mistaken . . .

Robert Frost,
"Blueberries"

What was your first experience with foraging?" The answer to this question will often tell you more about a forager's history than anything else you could possibly ask them. Lots of people have fond recollections of foraging with family members, shadowing them during the height of morel or fiddlehead season and being sworn to secrecy regarding the location. Or maybe as a kid they were introduced to blackcap raspberries by a friend in the neighborhood, away from the watchful eyes of potentially concerned parents.

Some introduced themselves to foraging—eating a ripe fruit from an unknown tree and sending their parents into a frenzy complete with an emergency room visit and frantic phone calls to poison control, only to discover that the offending organism was something benign, like a crabapple. Still others were taught to forage more intentionally, instructed on how to identify garden weeds by knowledgeable adults through the course of shared chores. Perhaps some were even initiated into the lifestyle by learning how to find the answers themselves, guided by some adult with a library card who didn't mind not knowing everything right away.

Many foragers come to the hunt later in life, as evidenced by the foraging classes I teach, which tend to be chock-full of eager adult beginners. Some of them have never so much as visited a U-pick apple farm, others may have a vested interest in gardening or local produce, and still others are seeking the next logical step in a journey already begun toward a more sustainable lifestyle. Plenty of them have half-remembered experiences mirroring those listed above, but have a long-held assumption that food from the wild is suspicious; only food from the store can be trusted.

Surprising to some (I am, after all, a *mushroom* auntie), my first wild food experience was actually with a plant—*Vaccinium corymbosum*, the highbush blueberry. It remains one of my most intimate relationships to date, spanning nearly three decades of patient teachings. For a full fifteen years before I ever plucked a wild mushroom, I was a berry picker—preoccupied with blueberries and wintergreen berries, sweet *Rubus* species like blackcap raspberries and blackberries, counting down the minutes until school or chores ended before I could run outside and begin collecting.

JULY 1996
Milford, Pennsylvania

Pay attention," my mother admonishes me, "he's trying to show you something."

My family and I are recent transplants from New York state, freshly moved into a new house in the Poconos, tucked down into a slope where an oak-hickory forest meets a vast shrub swamp. Our new neighbor Steve is a burly man with a handlebar mustache, a Harley-Davidson, and, fortunately for us, a soft spot for the neighborhood kids (occasionally he would even grant us the opportunity to ride around the block on the motorcycle). He points again at a short, scrawny-looking tree with papery bark. "That's a blueberry bush. You have lots of them around here on this side, but they're all over the community too, especially up by the road." He reaches his stubby yellow fingers out and plucks a tiny berry from the tree, handing it to me with a friendly smile. "My kids used to pick them. Dunno how much longer they'll be here, with the houses going in, so enjoy them while you can."

I examine the berry with skepticism. I love fruit, but if I remember correctly, blueberries taste like wet, mealy little balls of vague sweetness, indistinctive and unsatisfying. This looks a *little* like the bloated things that come in a plastic package from the grocery store, sporting the same little star-shaped flaps and the purplish-blue color, but it's much smaller and rounder. Not wanting to be rude, I pop it in my mouth, expecting to be disappointed.

Whoa. A sharp, sweet, tangy flavor erupts across my tongue. "*THAT'S* a blueberry?!" I exclaim, immediately scanning the tree. "Can I have more?"

"I think you've created a monster," my mom quips. She and Steve laugh as I bend the branches down toward me, reaching for grubby fistfuls of perfect berries and stuffing them into my mouth as fast as I could manage.

She had no idea.

BLUEBERRY PICKING OCCUPIED my every thought after that. I followed the trees, always barefoot, mapping out all the best spots up by the road where the full sun shone on them for maximum berry production. I poked holes in Tupperware containers and ran a string through them to free my hands—one to bend the tree down where I could reach the berries, and the other to pick. When I went to sleep at night, images of blueberries were stamped behind my eyelids, imprinted on my brain.

My love for blueberries was a child's love—curious, open, impatient, clumsy. The trees tolerated my intrusive prodding and constant examinations, my occasional rough treatment resulting in a broken branch here and there (always ending in shame and tears). I watched the deer eating blueberries, arching their slender necks like giraffes and eating directly from the tree, guided by their long tongues, and then I tried it too. Many have implied over the years that I was a feral child, and in some ways, I suppose I was, and maybe even still am (homeschooling probably set me up for it). My family jokes that I didn't wear shoes until I was a teenager, and even in my thirties I still feel most myself with scraped knees, sticks in my hair, walnuts in my pockets, and soil under my fingernails.

The blueberry bushes have grown with me, and have given me many teachings. Some have been practical, like the importance of balance—too much sun and not enough water and they will dry up before ripening; the opposite and they will rot. Treating the bushes with disrespect or carelessness, breaking their branches will cause them to go into recovery mode, producing fewer berries the following year while they concentrate their energy on trying to repair the damage that has been done. Other lessons have been deeper, reflecting the practice of patience, of tending to the garden of the forest, and of the undeniable personhood of the wild blueberry beings. Each one has its own characteristics, personality quirks that mean no two wild blueberry trees will produce exactly the same fruit.

Out in the brambly wood that would one day become my neighbor's lawn, just out of sight of my green front door, I met a grove of bushes, each dripping with enormous berries, the gold standard for all the bushes on my street. Some had lots of tiny berries, others had a few large ones. A few were seedy and tart, others burst with juice the moment they touched your tongue like popping bubbles in boba. All of them were good, but for whatever reason, these bushes in particular had hit the environmental jackpot, blessing me with annual berry perfection.

JULY 1999
Milford, Pennsylvania

After weeks of waiting, I woke up on a perfect July morning and knew it was finally time. They were ready, I was eight years old, and I had no time to waste.

I raced through the morning's necessities and dashed out the door with a plastic container strung with yarn dangling from my neck. The bottoms of my bare feet pounded on the gravel as I ran up the driveway, stained with the grass and dirt of the summer days already behind me.

They were perfect. The biggest wild blueberries I'd ever seen, heavy and dark beneath a milky-blue layer of bloom. So many that they bent the branches, like grapes. Somehow in a single day they had become even more lush and beautiful, ripening from purple-pink to perfection overnight.

Pure delight ran up through my body like sunshine as I wriggled from head to toe. All the waiting had been worth it. I started picking with both hands, barely needing to pull the branches toward me to reach them. Cascades of little *thwump* sounds marked their descent into my bucket, which pulled the scratchy yarn deeper into the skin on the back of my neck as it filled. Mosquitoes swarmed and drank from my arms and legs, but I barely noticed. At least, until I heard the sound of wood cracking just behind me.

I turned around, preparing to snap at one of my brothers for sneaking up on me, and was met with an enormous body, easily four hundred pounds, covered in black fur, so close that I could smell its hot breath.

A black bear.

Since moving to the mountains and deciding that I belonged outside as much as possible, I had been warned by the adults in my life of the dangers of the cottonmouth snake, which apparently liked to rest beneath the blueberry bushes and bite the toes of little girls who refused to wear shoes. I had also been warned not to touch mushrooms, which I was told have a nasty habit of leaping out of the ground and jumping down the throats of unsuspecting berry pickers. I know for a fact that I had also been warned about bears, the original berry-pickers of the Pocono Mountains. I had been given sound advice—that black bears are typically quite fearful of humans, that you should make yourself as large and loud as possible to frighten them off, but in that instant, I forgot everything I knew.

I just stood there, as still as a monument, staring with wild, helpless eyes as the bear feasted, not knowing what else to do.

His tongue was purple, and his paws looked sticky. It would have been comical if I had not been inches away from him, his fur practically grazing my knees as he waddled in front of me. "Tubby" is the wrong word to use. This guy could have flattened a Subaru. Looking back, it was impressive how well he had done for himself, and I sincerely hope that he had a very comfortable hibernation that winter, but in that moment, as he stood up on his hind legs and stretched out to his full height to reach the highest branches, I was terrified.

As I stood for what felt like hours, heart thumping in my ears, my fear began to give way to indignation. How *dare* this bear come in and eat all *my* blueberries!? I had waited for months, coming to visit them constantly, willing them to ripen, and here he was, eating them all by himself like he owned the place!

It was clear which of us had brought the bigger bucket, and it certainly wasn't me.

The whole time, and even years later, I still can't be sure if the bear actually saw me or not. He didn't seem concerned either way, instead choosing to preoccupy himself by stuffing his face full of fruit. I suppose I wasn't exactly intimidating, a skinny eight-year-old with a bucket on a string and twigs in my hair.

I suppose it's possible that I neither frightened nor intimidated him. The bear belonged to the forest in a way that I had never quite managed, a grandson of thousands of generations of bears that roamed this mountain back when it was taller than the Rockies, long before the roads were cut and the houses built. His ancestors' bones crumbled in the sediment beneath our feet, and I was simply a little girl with frizzy hair and a sticky, purple mouth, a child whose family had sailed across the ocean to land on these shores mere decades ago.

The bear remembered the song of the blueberries, of gifts and plenty, from all the blueberry seasons of his life. He ate, content with what he was given, unconcerned that I might encroach on his table, unlike my anxiety over what would be left for me. I was still being taught, a newborn in the forest, with much to learn and even more to unlearn.

I am still, as we speak, learning this song, clumsy and off-key despite years of practice. The language is difficult and deep, and the way is not easy.

YOU CANNOT OWN a forest, because you belong to it. Greed is a vortex; it spins the world out of balance and into darkness, consuming and consuming until there is nothing in your hands but dust where you once grasped at abundance.

You cannot own a forest. You can only love it.

Many North American Indigenous people call blueberries "star berries" in their own languages, remembering an ancient story where the Great Spirit sent blueberries to save the people from a terrible famine. The five points of the calyx—the blossom end of the berry—are reminiscent of a five-point star, and a mark of their status as a gift from the heavens. Settlers have shorter memories, but records show that the first pilgrims were similarly saved from starvation by blueberries given to them by the Wampanoag—a gift received and given again. Samuel de Champlain wrote that dried blueberries provided "manna in winter" when other food was unavailable.

A gift.

Freely given, but conditionally. The gift must be given, but it can only be received in relationship—care begets care. Hands that accept such a gift must be willing to love the giver back, or eventually there will be nothing left to take.

The seesaw of reciprocity is the quiet and constant assurance of future abundance.

We give our gifts with an equal measure of generosity in which gifts have been given to us, continuing the cycle so that we may hold one another up.

We give our gifts so that our hungry world can be fed.

AS I WATCHED the bear basking in the glorious summer heat, juices dripping from his muzzle, I felt ashamed, knowing I deserved to be scolded.

I did not own the berries, any more than I owned the bushes they grew on or the earth they came from.

You cannot own a forest.

You can only accept its gifts when they are given.

The bear turned its back to me, its interest shifting to another bush. In slow motion, I lifted my trembling hands and removed the bucket from my neck. It was heavier than it had seemed a moment before.

I lowered the string until the bucket rested silently on the sphagnum moss.

Quietly, with a deep breath, I turned my body toward home and ran as fast as I could, leaping down the hill in enormous strides that felt like flying. The blackberry brambles, sterner teachers than blueberries but no less effective (berries are generally worth the risk of wounds), left thorns like daggers in the bottoms of my feet. I did not look back until the front door was closed behind me. When my mother saw me, white-faced and terrified, I could only choke out one word: "*Bear.*"

I told her the story, breathless and disorganized. We looked out the window to check, but the bear hadn't followed me—blueberry girls don't taste nearly as good as blueberries all by themselves, after all.

A few hours later, when a neighbor farther down the road confirmed that the bear had moved along, I took my mother back to the place where it had happened. We saw the nest he had pressed unintentionally in the ferns, a veritable ursine crop circle.

Just outside the circle, I saw my bucket, three-quarters of the way full of berries, just as I had left it. The bear hadn't touched it.

Even the bushes the bear had visited still appeared full and lush. The bear had left plenty behind before toddling away, more than enough to fill up the rest of my bucket.

Together, Mom and I finished the job as I recounted the story again, pointing to where I had been standing and where the bear had been eating berries. She gasped in renewed horror when she realized just how close we had been to one another.

JULY 2022
United Plant Savers, Ohio

"I just found a *RIPE MAYAPPLE,*" I crow at the camera, buzzing with glee.

I'm in southern Ohio at United Plant Savers, working for a couple of weeks as an artist in residence at the Deep Ecology program. After obtaining special permission to forage (modestly) on this normally strict no-foraging property, I am alone in a yurt for the next ten days putting together my project—an ambitious combination of my various passions culminating in a foraged nine-course gourmet meal (a few store-bought ingredients are begrudgingly allowed), plus an evening-length soundtrack (created from field recordings collected onsite during my brief stay) served to local participants. No big deal. Normal, even.

Trying to explain to other people what I do is a bit of a challenge, to say the least. How do you tell people that you're a professional composer and visual artist, but your *other* job is to pick things up off the ground and teach people how to eat them? How do I explain that the part of me that needs to make art is the same part of me that needs to commune with mushrooms and trees? To me, the act of creation is no different in either case. I need both. I can't *choose.* They feel the same to me. When I have nothing left to give to my art, the forest gives ideas back to me. When I can't decide what to be, I go to the forest. It always knows who I am, all of me, and it never makes me decide between my left or my right.

WHEN I AUDITIONED to study music at Butler back in 2010, I met with one of my future composition professors (and current dear friend) Dr. Frank Felice, a warm man with wild gray curls down his back, a goatee, and twinkling eyes revealing a Gandalf-like wit and whimsy. If you met him on the street, Frank could be anything, but especially three things: a wizard, a rock star, and a composer. Realistically, he is all three, and more. His office overflows with the eclectic. Pop culture figurines, books about paleontology, scores, CDs, and cassette tapes to the ceiling, paintings, and a crispy flagon of Diet Mountain Dew. I'm proud to have a few of my own artworks and scores featured in his well-curated office these days.

"So, what do you want to be when you grow up?" he asked me, as if he were asking me what I thought of the weather.

I thought for a moment, surveying his office. "I think I want to be a polymath."

His laugh boomed, echoing around the office and down the hall. "I think we are going to be *very* good friends."

ALLIUM TRICOCCUM
Ramps, wild leeks
AMANITA MUSCARIA
VAR. GUESSOWII
Yellow fly agaric

AMELANCHIER SP.

Juneberry, shadberry, serviceberry, saskatoon berry

VACCINIUM CORYMBOSUM

Highbush blueberry

ARMILLARIA MELLEA
Ringed honey mushroom

DESARMILLARIA CAESPITOSA
Ringless honey mushroom

CANTHARELLUS SPP.
Chanterelle

CERIOPORUS SQUAMOSUS
Dryad's saddle, pheasant back

ASIMINA TRILOBA

Pawpaw

EXSUDOPORUS FROSTII

Frost's bolete

BOLETUS EDULIS CLADE

Porcini

COLLYBIA NUDA

Blewit

CORNUS MAS
Cornelian cherry,
Carnelian cherry
CASTANEA DENTATA
American chestnut
DIOSPYROS VIRGINIANA
American persimmon

GRIFOLA FRONDOSA

Hen of the woods, maitake

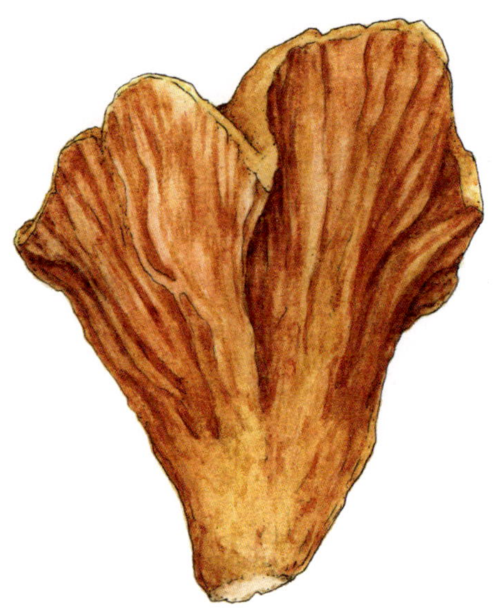

HYPOMYCES LACTIFLUORUM
Lobster mushroom

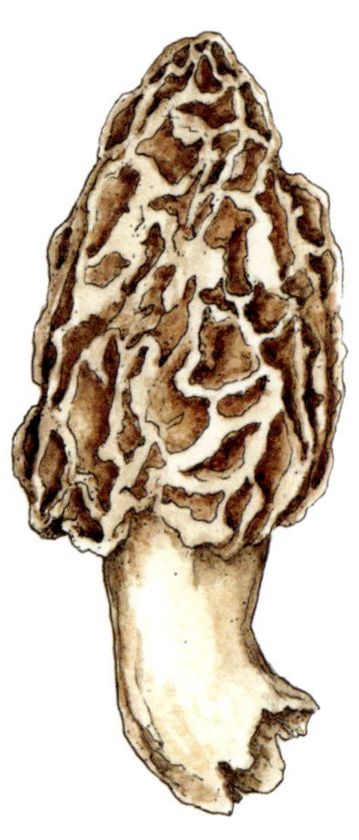

MORCHELLA AMERICANA
Yellow morel, American morel

FLAMMULINA VELUTIPES
Enoki, velvet foot, velvet shank

LINDERA BENZOIN
Spicebush
MAGNOLIA X
SOULANGEANA
Saucer magnolia

LACTARIUS INDIGO
Indigo milkcap
PICEA ABIES
Norway spruce
MONARDA FISTULOSA
Bee balm, wild oregano

PLEUROTUS
CITRINOPILEATUS
Golden oyster mushroom
PODOPHYLLUM PELTATUM
Mayapple

RUDBECKIA LACINIATA
Cutleaf coneflower, gray-headed coneflower, sochan
TYPHA LATIFOLIA
Cattail

RUBUS OCCIDENTALIS
Black raspberry, blackcap

RUBUS ALLEGHENIENSIS
Allegheny blackberry

SAMBUCUS NIGRA
Black elderberry, elderflower

MYOSOTIS ARVENSIS
Forget-me-not

HOUSTONIA CAERULEA
Bluet

VIOLA ODORATA
Sweet violet

SYRINGA VULGARIS
Common lilac
ROSA RUGOSA
Rugosa rose

LAETIPORUS CINCINNATUS

White-pored chicken of the woods

LAETIPORUS SULPHUREUS

Sulfur shelf, chicken of the woods

At the same institution, a few years later, a different professor cautioned me that I needed to choose—to devote all my creative energy to one discipline in the pursuit of mastery. "Look at Schoenberg. He was a good painter, but he knew he'd never be a *great* painter or a great composer if he tried to do both. Polymaths are the exception, not the rule." Greatness, to many of those deemed great within their field, can only be won through single-mindedness, a dogged determination to conquer your past self and push off any competitors that might be waiting in the wings to steal your spotlight.

"How can I do it?" I ask Lisa in one of our lessons, which seem to double as therapy sessions as of late. "How can I make it all work together? How can I be taken seriously by other composers when I don't care about the same things they do?"

She pauses, considering what I've said. "Are *you* embarrassed of what you do?"

The question catches me off guard. "Not exactly. I'm more embarrassed by the connotation, by what it makes other people assume about me."

"What do they assume?" she presses.

"That I'm a hack? That the things I care about don't really matter? That I don't fit into their world, their values, their academia?"

"Good!" she exclaims. "If the thing you want to be doesn't exist, then you'll have to be the one to *make* it exist. Build it from scratch. Be the first. *Make* them respect you. Prove that it matters."

So I'm here at United Plant Savers, purportedly both artist and naturalist, chef and composer, trying to do everything all at once.

As if the challenge isn't difficult enough, I decided to make it vegan and gluten free, just in case.

Easy.

I don't know if I have enough time to pull this off.

The best way to start is a walk in the woods.

I set off two hours ago with a basket, a Zoom audio recorder, and a few ideas of what I might encounter. Some culprits seem fairly likely—casual roadside invasives like chicory (*Cichorium intybus*) and dock seeds (*Rumex* spp.) are always plentiful if you care to find

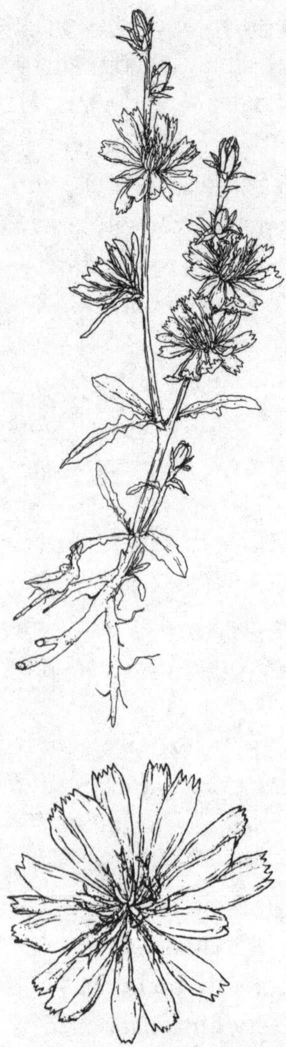

Cichorium intybus, Chicory

them, maybe a couple of mushrooms. The ripe mayapples are the first fruits I have seen so far, but they seem to be a sign of what is to come.

Mayapples, *Podophyllum peltatum*, also known as the American mandrake, are a common sight in forests like this, but their fruits are relatively rare to find fully ripe, as squirrels and other small mammals have a significant advantage when it comes to foraging them. Earlier in the season, large, umbrella-shaped leaves shelter a white flower that eventually grows into a yellow lemon-shaped fruit. They smell wonderfully sweet, like passion fruit and summer grapes, and they taste even better. However, aside from the ripe fruit, the rest of the plant is extremely toxic—even the seeds. Like many wild foods, careful processing is the key to ensuring a safe harvest.

Southern Ohio is long overdue for some genuine praise. There is a warmth and coziness to its hills, its grassy knolls and knobby glades, tall, stout-bellied trees, and creeks twisting spider-like through antique farms. Stunning natural features—gorges, waterfalls, and the staggering biodiversity that can be witnessed in the mixed mesophytic forests typical of central Appalachian Ohio. It feels like the Shire, as much as a place not *actually* populated by hobbits really can.

There is a word used to describe this type of forest—a *refugium*. Quite literally, a refuge—a habitat that provides a specific set of ecological conditions, allowing remnant species that may not be able to live anywhere else in the world some sort of resistance. I once heard the ethnobotanist Kathleen Harrison use the term when speaking about her mycological research among the Mazatec people, whose mountains harbor dozens of species of "magic" psilocybe mushrooms that can be found nowhere else in the world.

Refugia protect the most fragile among us from harm.

Something about the ground here seems perpetually damp. It hasn't rained terribly recently, but there is green everywhere I look.

Deeper along the path, wild ginger creates a thick mat on the forest floor, sending its rhizomes shooting out along the earth like soldiers sent to conquer more territory. Its fuzzy flowers send out a putrid, rotting-corpse smell, a bespoke calling card for the black flies that have evolved to pollinate them. In the available spaces between the colonies of ginger, deceptively friendly red leaflets of poison ivy, *Toxicodendron radicans*, pop out, glossy and serrated.

The forest suddenly opens up before me, and the tangle of brush, vines, and hardwoods becomes a sea of large white pine, seemingly out of nowhere.

It is easier to see now, between the tidy softness of the pine duff and the wide spacing of the trees. Mushrooms I could never have noticed in the hardwoods without diligent and thorough inspection

***Rumex crispus*, Curly Dock**

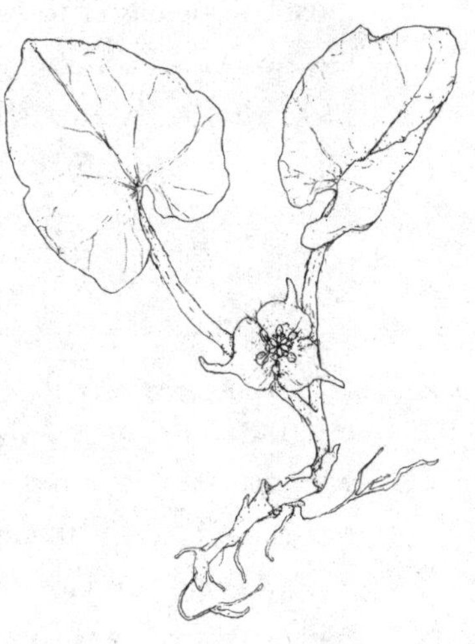

***Asarum canadense*,**
Wild Ginger, Canada Ginger

are suddenly visible with almost no effort at all. My eyes fix greedily on a flush of candy-apple red Frost's boletes, *Exsudoporus frostii*.

They appear almost glossy in the humidity, and upon further inspection, gemlike drops of golden guttation cling to the crimson hymenia. Observing the dramatic reticulation along the stipe never fails to delight, nor does the immediate blue bruising at even my gentlest touch, appearing nearly black upon the red flesh. I lick the top of one, enjoying the distinct lemony flavor of this species.

I collect seven or eight pristine specimens, carving away at soil and insect predation and revealing a yellow interior at the base of each mushroom. This, too, immediately begins to turn blue as the acids oxidize.

Other mushrooms also come into view from my vantage point—loads of *Lactifluus piperatus*, a latex-producing mushroom so incredibly spicy that it makes my eyes water just remembering the last time I tasted one, various *Cortinarius* fungi sporting an array of purples ranging from lilac to royal, dozens of *Amanita*, caps domed and flat, warty and smooth, plain and showy, and even some scattered *Mycena*, tiny but fascinating.

The pine duff also hides mushrooms, even as it reveals them, as mushrooms either too short or too young to fully emerge appear as little bumps, forming what myco-folks call "shrumps." I see a few spots that make me "shrum"-spicious. The spent needles are themselves a sort of refugium, a hiding place for shy fungi.

I'm so focused on finding mushrooms that I almost forget about the audio recorder.

Headphones on, *click* to record, and we're rolling.

I lean back on my heels, trying to be as silent as possible. The microphones are sensitive, and even my breathing can disrupt the soundscape. Listening to the amplified recorder through headphones has the effect of making me more aware of the sounds around me, even though fundamentally, nothing has changed since pressing the button. Birds are the most obvious sound at the forefront. I've always been terrible with bird identification, but I can pick out a few—a finch, sparrows, a crow. Creaking, shuffling, the *scritch-scratch* of small mammals foraging is next. When I listen very closely, I can hear spent pine needles hit the forest floor, like pins dropping.

When I'm satisfied that the baseline recording sounds good, I take off the headphones and walk away, letting it continue rolling on its own. I begin inspecting shrumps for edible fungi, carefully lifting the duff with my fingertips to see what is hiding under each one. Porcini, *Boletus edulis* clade, are prominent in my hopes, however dim a possibility. Foragers in the western United States and in Europe have the superior specimens, even the pins appearing large, sporting inflated stipes and delicate reticulation, with fat, doughy sponges plunked so precariously on top that they look like overlarge old-timey paperboy caps. Here, east of the Rockies, the porcini are stunted, often so thoroughly invaded by insects by the time they emerge from the soil that they are barely recognizable as porcini. I don't expect to find them, but it doesn't stop me from looking.

Some of what I uncover is a bit disappointing—rocks, pine cones, and buried branches love pretending to be mushrooms, teasing me. Others are beautiful specimens, *Cortinarius* and nondescript milky mushrooms, but none are edible or choice enough for my table.

The soundscape is peaceful, but not satisfying when I check the recorder with my headphones. Something is missing. I attach a terraphone to the recorder and dig down about six inches through the pine needles until I reach the soil, bury it, and listen again.

This is different, though no less lively. The birds and mammals are muffled, background now rather than foreground. I can hear gurgling, shuffling, movement. Tectonic sounds fill my headphones, bumping and knocking around like sonar. In such a strange auditory environment, there are imaginary, possibly microscopic sounds as well—I can picture the sound of mushrooms growing stout in the darkness beneath the duff.

Mushrooms sometimes seem to appear virtually overnight, though this of course is not because they simply choose to pop up one morning on a whim. For a mushroom to appear in one's lawn, a long list of environmental conditions must be met, and other processes must become involved; massive networks of connections like neurons must fire, collect data, and agree with one another. It takes a *ton* of energy to produce mushrooms, the fruit of mycelium. Mushrooms must absorb water—often huge amounts of it—from their environments, force themselves through challenging barriers like soil and wood, and then attract hungry creatures to carry them elsewhere.

When I explained this to my brother Christian, he immediately asked, "Why bother going through all that effort when you could just keep growing bigger and more powerful underground?" Fair question. Ultimately, there are two answers. The first is that evolution compels organisms to improve themselves by producing future generations, and the second is that nature operates in an economy of gift-giving.

Mushrooms are gifts from the fungi that produce them—they are given for free to whomever will choose to eat them, and unbeknownst to the eater, they are helping the fungus through their gathering. As a squirrel, an insect, a turtle, or even a human carries their prize through the forest, they are simultaneously seeding the forest floor with spores. It is pollination, of a sort. This ensures that the fungus can reproduce, can be replanted in places where, normally, only feet and wings can travel.

I continue unearthing shrumps, lifting the duff again and again only to replace it in ever-growing disappointment. Every mushroom is a good mushroom, and I always enjoy finding them, even if I can't eat or use them, but time is of the essence and I'm beginning to feel the crunch.

I lift another. *Lactarius* again. I've found about a hundred *Lactarius subpurpureus* so far. They're not terrible edibles when fermented, but not exactly what I'm in the mood to put on the dinner table either. I replace the duff, barely looking at the mushroom, then pause. I lift it again, then pluck the mushroom and turn it over. I already know what it is, but I can't believe I almost missed it.

A perfect *Lactarius indigo*, the indigo milkcap. Perhaps one of the most beautiful edible mushrooms found this side of the Rockies, its gills are the color of sapphire, a blue that makes the mushroom seem almost cartoonish and unnatural, like something that couldn't possibly belong in a forest at all. With shaking hands, I remove my knife from my basket and carefully slice the mushroom in half, watching blue latex practically gush out over my hands. The latex (as well as the rest of the mushroom) will eventually turn teal-green as it oxidizes. It is a good edible, a *very* good edible. The cap is occasionally a bit misleading to the naked eye, deeply depressed in the center with much lighter and more reserved shades of blue and gray alternating in concentric circles. The cap margins are rolled inward, resembling a cake donut, smelling nearly as sweet, and it is not until you turn it over that you can truly appreciate the color, much less marvel at how this bizarre mushroom could possibly be *food*.

Other shrumps near this one contain more and more of them, dozens of perfect specimens. I collect them, trimming them as carefully as I can to preserve as much latex as possible. As soon as I get them home, I will chop them up, squeeze them to release the latex, and infuse them in cream to preserve the color as much as I can for a later project.

Everyone else fades away for a moment—my colleagues, academia, expectations. I suddenly feel no urge to justify myself. It matters because it *does*. It matters because these mushrooms are the art I came here to see, from which will spring forth the art I came to make. It doesn't matter if anyone else approves; I know that this is *exactly* where I need to be and what I need to be doing.

This gift is just for me, today. My basket is full, a grateful remembrance of spores nestled into the altar of pine duff, stained blue like water, blue like sky. I listen in the headphones again for a few minutes, then click the recorder off and disconnect the terraphone, re-entering the familiar soundscape of the above-the-ground pine forest.

The earth is not quiet today, but it does seem to be full of holes where blue mushrooms once hid, proof of a curious underground society, one that needs no recognition. A society whose songs play in my headphones, entangled in a chorus of mixed voices, harmonizing in clicks and shimmers, drones and echoes.

Gathered, as one.

AUGUST 2022
Milford, Pennsylvania

"What was your first experience with foraging?" To be honest, I'm not sure my first experience ever really ended, because twenty years later, I am once again running outside on this promising summer morning, a plastic bucket bouncing on a string slung around my neck. I race through the trees, larger now, my bare feet bruising on the sharp slabs of lichenous rock jutting out from the earth, a contrast from the deep pillows of moss that surround them. I'm stubborn, unwilling to admit that my feet may be more tender now, as they are tasked with carrying a person significantly heavier than an eight-year-old.

The blueberry bushes still live outside this home, my childhood home. I left years ago, but we still recognize each other when I come to visit, always picking up where things were left off, even if it has been years.

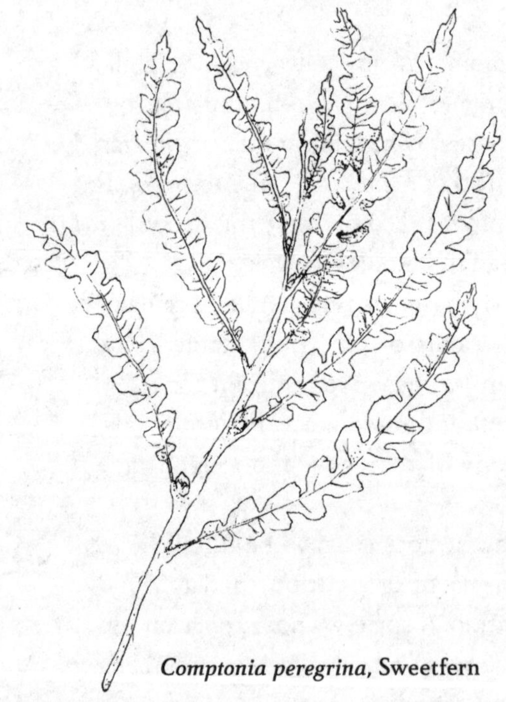

***Comptonia peregrina*, Sweetfern**

They are my grandmothers, long-suffering and wise to me as ever, even though I have surpassed many of them in height. No longer do I fit on the tiny platform treehouse I built in my childhood, armed with old nails, a borrowed hammer, and my dreams. Every time I go back, the blueberry bushes' numbers are diminished, their leggy bodies removed to make way for new housing, for driveways and lawns and swimming pools.

Still, I sit in perfumed beds of sweetfern and sphagnum moss and listen as those who remain recount the old stories. I hear them remarking on how much I have grown, and marveling at how little I have changed.

I still have so much to learn.

LISTEN. CAN YOU hear them? The shimmering percussion of trees, the tangled chorus of mushrooms. The songs that are sung beneath the earth, calling you to learn your part.

Become a part of it.

Here, you are already known.

GATHERING EXERCISE

"HYPHA"

- Find a place where the soil is soft
- Where the covering earth can be lifted with ease, fluffy in your hands
- Dig with your fingertips until you find a thread of mycelium, pliable and ivory-white like liquid paper dried on a rubber band
- Follow the thread for as long as you can
- Moving the soil around it gently
- As if trying not to wake a sleeping baby
- Follow it until it breaks
- Or it becomes too intricate to follow.

Wild Blueberry Ginger Cocktail Syrup and Gimlet

A NOTE ON WILD GINGER

While people have consumed wild ginger as a culinary herb for years without issue, it does *contain aristolochic acid, which is linked to kidney problems and certain cancers. It isn't recommended for regular use, and you should always check with your doctor if you have conditions that might be contraindicative. I personally feel comfortable using a small amount in a cocktail once in a blue moon, but everyone is different. If you don't want to use it, just sub in regular ginger.*

Wild blueberries and wild ginger go together like America and apple pie. The warm, earthy flavor of the ginger brings a sort of pan-seasonality to the acidic one-two punches of blueberry and lime. Combined, they make a syrup that will have everyone asking for a second cocktail. Because of the tart quality of this drink, I prefer to serve it on the rocks, since the ice keeps the acidic components extra bright and refreshing.

Contrary to what you may assume, when making a syrup with berries, frozen tends to work a little bit better than fresh. Have you ever noticed all the extra juice that collects at the bottom of a bag of frozen berries? That's because the freezing process expands the berries by forming interior ice crystals, which do the work of breaking the berries open for you, making the juices easier to access. I freeze my fresh berries by dumping them out on a baking tray in a single layer, then sticking the tray in the freezer for at least two hours. Once the berries are frozen, they can be removed from the trays and packed back into freezer bags for later use.

Pro tip: When you're done making the syrup, don't throw the strained blueberries and ginger away! You can toss them in a blender, make a paste, and turn it into fruit leather by spreading the paste out onto a silicone sheet and dehydrating at 120°F for 6 hours, or you can put them in a smoothie or use them to top ice cream.

To make the syrup:

Makes about 1 quart

2 cups granulated sugar or honey

About 2 cups blueberries, frozen then thawed until squishy

About 4 inches of wild ginger rhizomes, minced and cleaned (if you can't access wild ginger or don't want to use it, use 2 inches of regular ginger)

½ teaspoon citric acid (optional but will increase shelf life and enhance the flavor of the blueberries)

2 cups water

Mesh sieve or fine metal strainer

Cheesecloth

Place 1 cup of the sugar or honey in a saucepan and add the berries, wild ginger, and citric acid (if using). Smash the berries roughly into the sugar and mix together well, then let macerate at room temperature for about 10 minutes (this just means letting the berries and sugar hang out in the pan together so the berries soften up and the sugar absorbs the flavor). Add the remaining 1 cup sugar or honey and the water to the pan and heat over medium-low heat, stirring occasionally, until it comes to a controlled boil. When all the sugar has dissolved, 3 to 5 minutes, remove from the heat and let rest until fully cooled. If using honey, simply boil for 3 to 5 minutes.

Strain the mixture through a fine-mesh sieve, then once again through cheesecloth until you have a clean liquid.

The syrup can be stored in the refrigerator for up to 3 weeks. Do not store outside of the refrigerator, as small pieces of blueberry and ginger that escaped the sieve and cheesecloth are likely to mold if you do so.

To make the gimlet:

Makes 1 gimlet

Ice

2½ ounces gin

½ ounce freshly squeezed lime juice

½ ounce blueberry ginger syrup

Lime wedge or fresh mint, for garnish

(continued)

Fill a cocktail glass with ice. To a cocktail shaker, add enough ice to fill halfway, then add the gin, lime juice, and blueberry ginger syrup. Shake vigorously for about 30 seconds, then strain over the cocktail glass. Garnish with a lime wedge or a sprig of mint.

Blueberry Goat Cheese Galette with Sweetfern

Serves 4

I love making galettes. They look fancy, but they're the laziest pastry you can make. Whip one up for Saturday brunch or serve it up for dessert when you haven't planned ahead. On that note, you *can* make your own crust, but store-bought is also fine. Sweetfern has a pleasant, herbal earthiness that pairs beautifully with dairy ingredients, especially tangy cheeses like chevre. If you find sweetfern (*Comptonia peregrina*), don't forget to look around for blueberry bushes—I have fond memories of following the smell of sun-kissed sweetfern to find new blueberry picking spots.

For the crust:

2½ cups all-purpose flour, plus some for rolling

1 tablespoon salt

6 tablespoons unsalted butter, frozen and grated

⅔ cup vegetable shortening

½ cup ice water

2 tablespoons ice-cold vodka (optional)

1 egg (for egg wash)

For the filling:
4 ounces plain soft goat cheese
About 3 cups blueberries
About 3 to 4 leaves sweetfern, fresh or dried, finely chopped
Drizzle of honey, for finishing

Preheat the oven to 375°F and line either a pizza pan or a 10-inch cast-iron skillet with parchment paper. A galette is a relatively flat pastry, so while you may choose to use a pie pan, it isn't necessary.

First, make the crust. This recipe makes enough for two crusts, so you can freeze one for later whenever you need it, or you can choose to halve the recipe. Sift the flour and salt into a large bowl, then add the grated butter and shortening. Work together with two forks or in a stand mixer on medium-low speed (don't touch it with your hands, as you want to keep the dough as cold as possible) until the mixture resembles floury peas, then work in ice water and vodka (if using) a little bit at a time. The vodka keeps the dough moist and pliable but evaporates more quickly than water in the oven, which helps to develop a flakier crust, but you can skip it and just use a little extra water if you prefer.

Use your hands to form the dough into a ball, stopping as soon as it is formed. Don't overwork it! Cut the dough into two equal pieces. Wrap each piece separately in a towel or food-safe wrap and let it rest in the refrigerator for at least 30 minutes. Place one half of the dough in the freezer for later use, then turn out the other half of the dough onto a floured surface and roll out into a thin 12-inch circle.

In the center of the dough circle, add a layer of softened goat cheese, then arrange the berries on top. Add the sweetfern leaves. Fold the dough about 2 to 3 inches in toward the center like an envelope, then brush the edges with egg wash.

Bake for 30 to 40 minutes, until the crust is golden brown. Finish with a drizzle of honey and enjoy!

Mayapple Caramel Sauce

Makes 1 pint

For a long time, I believed (as many do) that mayapples never ripen off the plant. Unlike many fruits, mayapple will not typically ripen when placed in a paper bag with a banana or other ethylene-producing fruit. Instead, they can be ripened in a basket on your counter. Collect them when they are at least 50 percent yellow, then place them in a single layer in a basket. Mayapples ripen from the bottom up, so turn them over once a day like hatching eggs, arranging each one so that the ripest side faces down. A great tip from my friend JB is to cover them with a sheet of parchment while ripening so that they ripen in bright yet diffused light, which can prevent them from going too wrinkly. Depending on how green they are when you start, they should be ripe within five to ten days. You'll know they are ready to use when they are soft and smell strongly of passion fruit. The tart flavor of mayapple is excellent for tempering the occasionally cloying sweetness of caramel. I stir this sauce into my coffee, top my ice cream with it, and drizzle it on cakes, fruit, and even the odd cocktail. The butter and cream in this recipe can be substituted for vegan butter and coconut cream if you want a plant-based caramel.

Important note: *Only* the ripe fruit of the mayapple is edible. Every other part of the plant should be considered toxic, including the seeds *inside* the ripe fruit. Some people claim to ingest the seeds without issue. However, while I've seen plenty of excellent evidence demonstrating that the ripe fruit is nontoxic, I've never seen anything proving the safety of the seeds. Better safe than sorry.

½ pound ripe mayapples

1 cup granulated sugar

6 tablespoons unsalted butter, softened

½ cup heavy cream

1 tablespoon kosher salt

First, prepare the mayapple pulp: Slice your mayapples open and scrape out the pulp inside, then press the pulp through a sieve or mesh strainer with a spatula. Make sure no seeds are transferred through the sieve, as the seeds are toxic.

To make the caramel: In a heavy-bottomed saucepan, cook the sugar over low heat, stirring a few times per minute. It will begin to form clumps and then melt together. Watch the sugar closely, because it can go from "perfect" to "burnt" in a matter of seconds! When the sugar is starting to smell like toasted marshmallow and is beginning to turn from white to yellow, whisk all the butter in immediately. Once the butter has melted, add the mayapple pulp, cream, and salt all at once. Whisk everything together vigorously until the texture is smooth, thick, and syrupy, but still easy to stir, then remove from the heat and let cool for a few minutes in the pan. As the caramel cools, it will continue to thicken.

While the mixture is still warm and runny but not too hot to touch, pour into mason jars. Store in the refrigerator for up to 6 months.

Umami Bolete Dust

Ah, the humble bolete. Some of the friendliest entry-level fungi, generally characterized by a thick stem and a cap with a squashy, spongy, porous underside. There are very few dangerously toxic boletes, and many of the preferred edible varieties, such as porcini (*Boletus edulis* clade) have quite a distinct checklist of identifying features. Many boletes, while lovely to eat fresh, shine brightest after they have been dried. Some East Coast porcini hunters argue that dried boletes taste best when they're full of bugs (gross), but something tells me they're just saying that because they (and by *they*, I mean *me*) are enormously jealous of all the foragers in Colorado who casually pluck pristine, insect-less *Boletus rubriceps* from beneath the towering mountain conifers. Some of these mushrooms may rival the size of a small child, to my eternal annoyance. (Why couldn't it be *me*?!)

(continued)

However, where I live on the East Coast, what we lack in monster porcini, we make up for in sheer bolete species diversity. In Colorado, you might find fewer than twenty edible types of boletes, but throughout the Midwest and the mid-Atlantic, we have hundreds, if not thousands. One of my favorite things to do with edible boletes (especially if I have a large mix of species) is to make umami dust. The gentle heat of a dehydrator just *does* something to boletes—it concentrates all the best aspects of flavor. Bolete dust can be stored in a jar on your shelf almost indefinitely, and I find that I reach for it almost constantly whenever I need to pep up a boring pasta sauce, flavor a ramen, give a pizza some zip, or revive any number of disappointing kitchen experiments. I like to start with a quart and add to it over time. Make your own bolete dust and soon you'll understand why I have at least three jars of the stuff at any given time and it *still* doesn't feel like enough.

Edible boletes of any kind in any amount
A way to dehydrate (I like dehydrating large quantities of boletes on the dashboard of my car in the sun to save energy)
A desiccant packet (optional)

Slice and dehydrate your boletes. When the mushrooms are cracker-dry, put them in a spice grinder or a blender and blast them on the highest speed until they turn to a fine powder. Let the dust settle, and store in a mason jar with a desiccant packet, if you have one. You can add to your umami dust jar throughout the year as you find more mushrooms, or make different jars for each season.

Indigo Ice Cream

Makes 1 quart

I like the flavor of *Lactarius* mushrooms, but I can't say I'm a big fan of the mouthfeel. When cooked, *Lactarius* species tend to take on a somewhat gritty, grainy texture, not unlike a brick of cream cheese that has been frozen and then thawed back out. You won't see a *Lactarius indigo* on my plate unless it has been used to dye my ice cream blue.

There are a lot, and I mean a *LOT* of recipes for indigo milkcap ice cream. I certainly haven't reinvented the wheel with mine. When making indigo milkcap ice cream, the biggest challenge is finding a way to preserve the blue color without allowing it to oxidize, as oxidation will turn the mushrooms a far less appetizing shade of green.

How do we prevent oxidation? Science, of course! The fat molecules in milk and cream are excellent at slowing down the oxidation process, since it requires a *whole* lot of work to break down their structures. By keeping our dairy as cold as possible once we have combined it with the mushrooms, we can further delay this process. You can also "cheat" and add a bit of natural food coloring, such as blue spirulina or butterfly pea flower.

Indigo milkcaps don't have a ton of flavor on their own. Their flavor has been likened to "birthday cake," which I don't quite agree with, but I can absolutely identify a mild sweetness and a slightly buttery aftertaste in the flavor notes. You can add mix-ins or other flavors, but try it on its own first!

2 cups whole milk

1 teaspoon salt

¾ cup granulated sugar

1 tablespoon vanilla extract

2 cups heavy cream

10 to 15 indigo milkcaps, extremely fresh (store in the freezer until use or pick right before you need them)

1 tablespoon rum or vodka (optional, to keep ice cream scoopable)

1 tablespoon butterfly pea flower powder or blue spirulina (optional, for color enhancer)

(continued)

Combine the milk, salt, and sugar in a saucepan over low heat. Whisk either until the sugar is fully dissolved or steam is beginning to rise. Remove from the heat without letting the pot come to a full boil, then whisk in the vanilla and heavy cream. Pour into a mixing bowl with a lid and refrigerate for at least 1 hour. Add the vodka or rum to the ice cream base (if using).

Using clean hands, crush and squeeze the indigo milkcaps and drop them directly into the ice cream base. Work quickly to avoid oxidation. Stir the milkcaps in with the butterfly pea flower or blue spirulina, if desired, cover the mixture with cellophane or beeswax wrap pressed down to the level of the mixture, then place in the freezer for 30 minutes to infuse. Remove from the freezer and strain all the mushroom chunks out using a colander or a wire strainer.

If you have an ice cream machine, churn immediately according to your machine's directions. If you do not have an ice cream machine, place in a low, wide container with a lid in the freezer and stir every 20 minutes, scraping the bottom and sides thoroughly until it reaches a scoopable, velvety texture.

The Land Is in You

Open your mind.
Get off the couch.
Move.

Anthony Bourdain

I became acquainted with the work of Anthony Bourdain approximately two weeks after his death in 2018, when a foodie friend suggested that I might agree with Bourdain's opinion on Icelandic cuisine. As it turns out, I did agree—despite living there for a while and loving the place, I never got past Icelanders' cultural inclination to make both licorice and fermented fish taste like pure ammonia.

Despite never knowing him in life, or even knowing *of* him in life, the more of his work that I consumed, the more I found myself mourning his loss. He was a grumpy old man with a stubborn streak that probably meant that working with him was a real pain in the ass. A notorious bad boy chef in his heyday, he admitted in various publications to being practically tyrannical in the kitchen, painting himself at times as a cartoonishly angry, almost Gestapo-like character who engaged in all the vices of the '90s, from casual sexism to doing blow in the walk-in (probably). But in his later, more sober years, he found a capacity for something deeper, quieter—becoming the rarest type of American, one with the cultural sense to know when to shut his hole, and the openness to participate in the act of cultural exchange without patronizing or exoticizing the people he visited.

As a forager, I relate to certain aspects of Tony's mission. We're far from identical (I'm not an obligate carnivore, for starters), but in many ways, we have both taken up the mantle of persuading people to try foods that they often find scary or disgusting, usually at a table seated alongside people that they may have complicated feelings about.

Something I'm constantly reminded of when I watch reruns of *No Reservations* or *Parts Unknown* is the fact that food is always, *always* political. Every choice any of us makes about our food is affected by local and global economic structures, regulatory processes, international relations, and trade. The bodies of those who design, harvest, prepare, process, package, and serve our food are politicized, whether they want to be or not. Food, or lack thereof, can serve as a way to control entire populations, or, conversely, be employed as an olive branch to unite people, allowing us to see ourselves in our enemies. Food speaks louder than words, and that's what makes it so revolutionary (and often, so disruptive).

To many, foraging may seem like a way to escape all those political complications. *Free food*, divorced from any government overreach, practically guiltless! To others, foraging is a selfish act, nothing more than an inconsiderate theft of resources that belong not to humans, but to the animals that are solely dependent on them.

Ultimately, both perspectives fail to see the whole picture. Nature is not a domain to be conquered, a dense collection of dumb and unfeeling plants, animals, and fungi for our unfettered use. Nature is also not a utopia that could be *perfect* if those pesky humans would just quit involving themselves. Geographer William Denevan referred to this as the "pristine myth," the belief that all of nature was an untouched wilderness, over which humans had no influence and left no mark. Indeed, there are very few places on Earth that have not been marked by the ecological engineering of human beings.

This lie, the assumption that the removal of humanity is the key to a thriving ecosystem, has been responsible for the continuing erasure of millions of Indigenous people from the heartbeat of the lands they have shaped since time immemorial. We stole children and shut them away in boarding schools, forced them to speak our languages and forget their ways, but like a plant hidden in a closet, knowledge will drink in whatever light it can find, even if it is fleeting, barely there, a crack in the door. Even when weakened by the crushing blows of colonization, neither human nor land ever fully yields to it. We have to reckon with it. We have to own it.

We know that humans have trod in the footprints of even the most ancient of creatures and have shaped the land with fire, earth, and water to support life. We have also seen scars carved into the earth by our bombs, watched our pollution choke the waters. As much as we hate to admit it, this too is nature. We fickle, warring, innovative creatures are just as much nature as the deer foraging acorns in the suburbs, the lone orchid growing in a secluded wetland, or the Grand Canyon carved away by the ravages of time and water. We're stuck with each other, like it or not.

The truth, as with most things, lies somewhere in the nuance, a tenuous balance between giving and taking. Like all relationships, the conversation between human and nonhuman is a negotiation, taking into consideration a multitude of factors like abundance, one's own need, and the harm that may or may not be done to either the organism or the environment in the process of harvesting.

If I take something, what am I doing to return the favor?

How can this transaction leave both parties better off than we started?

This is the geopolitical reality of food writ small.

To my knowledge, Tony wasn't a habitual forager, though with all his experience and sensibilities I think he could have appreciated the act. (However, I'd probably have to insist that he refrain from lighting up a Marlboro in the forest, which may have been a dealbreaker.) He certainly didn't shy away from adventure or from a good meal, and if you could combine the two, well that's just a bonus. It doesn't get a lot more adventurous than hands-in-the-dirt, might-kill-you-if-you're-not-careful foraging for wild food. Kinda makes you wonder if foragers, not chefs, are the real punk rockers of the food world. (Shots fired!)

Like many of Tony's contemporaries, unwashed characters abound in my own foraging history. In a subculture defined by the acquisition of and enthusiasm for (basically) free food, you'd be surprised at how many people hoard their knowledge. Grudges can sometimes be held for years, often originating over something small like a secret location that turned out to be not-so-secret, or even a methodical or speciation disagreement. Without ever once intending to cause a stir (I hate conflict), I've had my life and safety threatened by people online who recognized a public spot where I was gathering, demanding that I cease and desist because that was "*their* spot." The concept of ownership, be it information, locations, or connections, is a baffling one, persistent in what should, by all appearances, be a rather socialist lifestyle.

On the other hand, those experiences are easily washed away by some of the kindest and most generous people I know, people who I am grateful to count among my dearest friends.

It seems that those who cling to the idea of such ownership have neglected something very important in their relationship with place. They have forgotten the deep magic, the whispered symphony sung with all voices, human and not. As Ojibwe activist Winona LaDuke reminds us, "Food has a culture. It has a history. It has a story. It has relationships."

Humanity is carried upon this magnificent earth, not as parasites, but as symbiotes. We cannot truly own it in a capitalistic sense, just as we cannot own each other—we are instead a part of it, as closely entwined with its fate as fungi is tangled within the cellular structure of the trees.

And indeed, the patterns of our own flesh.

We are at once owned and owner, belonging and belonger.

We are, each of us, held in common by the earth we share, far more than our society can possibly understand.

Whether we like it or not, we are in the land.

The land is in us.

SEPTEMBER 2015
Holliday Park, Indianapolis

I've recently taken up trail running at a local park, a sport at which I am unquestionably terrible. My mother always apologizes for passing the "klutzy" gene, which I frequently blame for my poor performance in sports (certainly not the fact that I could usually be found picking flowers in right field while balls sailed past me or, occasionally, into my inattentive face).

But it's okay to be bad at running. Trail running gets me out of the house and makes me feel as though I've been doing something productive with my body, plus it gets me some peace and quiet.

At this point in my life, a year and a half removed from the imposed structure of university life without much of a plan for moving forward, I desperately need time to think. Time to figure out my next move, to make the world seem manageable again. *What was all that work for? Come to think of it,* who *was it for?*

The park has constant elevation changes, with winding staircases of slippery wooden planks smoothed and worn down by the perpetual caresses of human hands and feet, made more treacherous by recent rain. I pant and struggle. The banality of performing my current job as a desk jockey has left me completely out of shape.

I take a moment, leaning against a railing. The sudden cooling of the air coupled with the unfamiliar feeling of exercise makes it hard to breathe. *Yikes. Kinda pathetic.* The days are still warm, hot at times, but I'm not accustomed to the growing chill in the evenings yet. The smell of petrichor and the funk of decay surround me, organic and earthy.

I briefly consider burying myself in a pile of damp leaves and taking a nap. However, such a position is rarely pleasant for long, and the cold would certainly catch up with me if I stopped moving for too long.

Just a minute longer, I tell myself, red-faced and panting.

My eye is drawn to a large clump of tan mushrooms to the left of the stairs. *Odd,* I think to myself. I've never seen mushrooms growing like *that* before. I think about it for a moment, and realize that I've actually never really *noticed* mushrooms before. Maybe if they were particularly colorful, or large, but they never registered with me for long.

As a child, I was told never to touch any mushrooms I found in the woods, since most of them were deadly toxic and it was difficult to tell the good ones from the bad ones—I wasn't one to negotiate with certain death. I remember hearing horror stories that usually involved someone's aunt's friend's neighbor's teenage son accidentally eating a mushroom from the yard and dying a horrible death mere moments later, such retellings typically replete with B-movie blood dripping from their twisted mouths before the ambulance could arrive. Even touching or breathing near a toxic mushroom could cause the spores to enter your bloodstream through the pores in your skin or your nose, for all we knew. So, we stayed away.

Something about this logic, while acceptable to a fairly gullible child, doesn't sit right with me as an adult. I've never thought to question it before, mostly because I'd never really thought much about mushrooms in all that time. I don't even really *like* mushrooms; wet, slimy things with vaguely meaty flavor and unnatural, chewy texture. A bit too buddy-buddy with death for my taste.

But that day on my run, a strange magnetism compels me to examine this clump of mushrooms a little more closely. Maybe it's time to look. They are brown, slightly tawny, all clustered tightly together, with a sort of purplish-brown edge on each cap. The cap centers seem just barely recessed, with a little raised "pop" up in each one like a nipple.

I take my phone out and snap a photo from where I stand, uncertain if I should actually touch them or not, and having no idea what kind of information that would glean anyway.

Plants were easy enough. I'd been learning about them since I was a child, their names and various uses practically tattooed on my brain. I could confidently and cheerfully survive for at least a little while on wild foraged vegetation if I needed to, happy to collect edible fruit, nuts, greens, and roots from many dozens of species I could recognize. My knack for plant identification helped ease me through more than a few thin paychecks in my time, though not everyone in my life felt confident enough in my knowledge to let me cook for them.

Mushrooms, on the other hand? Not a clue. I'd always assumed mushroom hunters had a death wish, that they just *really* enjoyed the feeling of living on the edge. The advice I'd been given was something akin to "don't mess with mushrooms, even experts end up dead every day." Somehow, in my child's mind, dying from a poisonous mushroom seemed like an even *worse* death than other kinds of poisoning, possibly just from the sheer embarrassment of it. While I might try my luck with wintergreen berries, you'd *never* catch me experimenting with mushrooms.

I've never experienced a feeling quite like this, but for the first time in my life, I am genuinely curious about the identity of the mushrooms in this strange cluster. It isn't a passing interest either—it is a compulsion, a crippling infatuation. I *must* know what they are. I am consumed with the gravity of this question.

Where do you even start with identifying mushrooms? I muse to myself. *Google? A book?*

My next stop is the library.

I check out the only three books on mushrooms that I can find: *The National Audubon Society's Field Guide to Mushrooms, Peterson Field Guide to Mushrooms of North America,* and *All That the Rain Promises and More,* the last of which I select less for its probability in helping me identify the mushroom and more because of the cheeky-looking man in a tuxedo with a trombone and a cluster of mushrooms on the cover.

When I get home, I settle in and try to use the keys in the book to identify the mushroom.

Almost immediately, I realize that in order to use a mushroom key, it is minimally imperative to know at least a few things about mushrooms. My plant knowledge doesn't really translate here, since I know nothing about mushroom morphology or terminology. New concepts began to overwhelm me. What is a hymenium? What is the difference between a gilled mushroom (and allies) and a polypore? Words like *saprotrophic, mycorrhizal, ectomycorrhizal, stipe, striations,* and *guttation* are tossed around with impunity as though someone like me should understand what they mean. Questions abound, like "How can I determine spore shape if the spores are microscopic? Do I have to buy a microscope to identify mushrooms?" and "How do I know if the gills are adnate or adnexed when I don't even know what the point of a gill even *is*?"

This is proving to be harder than I thought. Clearly, the books are too advanced for me at this stage in my knowledge-gathering. I push them to the side and open up my laptop to turn to the Internet.

Unsure of where to begin, I type "Indiana mushroom identification" into the search bar. The top result is a Facebook group with a few thousand members, simply titled "Indiana Mushrooms." The tagline reads: Document and identify mushroom species of Indiana. Post pictures of your latest finds.

Wow, I think to myself, breathing a sigh of relief. *Jackpot.* This is what I've been looking for.

I click on the page and answer the questions required to join the group, simple ones designed to weed out out-of-staters, a few basic questions about my knowledge base (zero), a disclaimer on psychedelics, a prohibition on sales and self-promotion, and a warning that everything you might choose to eat is at your own risk, and it is not the legal responsibility of moderators or group experts to prevent you from exercising your own brand of dumbassery.

Fair enough.

I'm accepted to the group within the hour.

I read through some of the posts, trying to get a sense of how this works. Do I just post a picture and get a response? Is it considered rude to start asking for help right away? Should I try to figure it out on my own by looking at other people's pictures first?

The answers are somewhat unclear. Some people seem more knowledgeable than others, using Latin binomial nomenclature in their posts and identifications. Other people use common names, commenting things like "COW" or "witch's butter." These people sometimes get ridiculed or scolded in the comments. Sometimes the people who seem more experienced will argue with each *other*, usually about the subtleties between one species or another. A few people's names surface more frequently, and many of these seem to be held in higher esteem than others. Some seem relatively kind, while others seem exasperated with apparent in-crowd social sins that I am growing increasingly fearful of committing. I learn through reading comments that you can touch mushrooms, even deadly toxic ones, which is a slightly scary but deeply exciting prospect. *I'm gonna touch everything*, I think to myself gleefully.

I'm nervous to post, but I *have* to know the answer to my question. I upload my photo, type up something about it that sounds marginally intelligent to me about what kind of environment it was found in, explain that I'm new to mushrooms but eager to learn, pause, add does anyone know what this mushroom is? and hit post.

Twenty minutes later, I have a response from one of the members, someone who seems like an expert. He uses a lot of Latin in his responses, which seems like a good sign.

Compare with Armillaria sp., possibly A. tabescens. Need underside pics to confirm. In future, camera needs to be closer to subject for ID purposes.

What is Armillaria? I'm elated. I snatch up the Audubon book from the dining room table and flip to the back, looking for any pages containing "Armillaria." I locate the page number for "*Armillaria tabescens*" and thumb my way to the entry.

I see that he's right. The pictures look similar, but my photo doesn't have nearly enough information to make a definitive identification. I can't see what the underside looks like, I can't tell if the stipe (which I learn is the fancy mycology word for *stem*) is indeed woody or not because they're not visible in the photo, I didn't take note of the nearby trees, and I also didn't get close enough to smell or touch them.

I also can't help but notice that according to this entry, "*Armillaria tabescens*" is supposedly edible.

I pocket the book and a paring knife and walk back to the park, darkness beginning to settle in around me. The forest looms darker still, but I push on, remembering the spot from my run.

The mushrooms are still there, just as I left them. I kneel beside them and use the flashlight on my phone to illuminate both the mushrooms and the page as I attempt to match features. I realize just how sensory the process is, tugging a single mushroom away from the bunch, examining the thready white strings of mycelium hanging off the bottom, inhaling the musky scent to determine if it does, in fact, smell faintly sweet. Peeling the stipe in two, I determine that it is in fact distinctly fibrous, and visual inspection of a few specimens rules out any skirtlike ring near the top. The caps are faintly tacky, a detail I hadn't noticed until I had reason to handle them, and some of the caps that had been stuck under other mushrooms in the bunch have been dusted with a whitish powder, which I later learn is a natural spore print.

Everything seems to check out.

I realize that I haven't brought a bag or a basket. I look around (as if a basket would magically appear for me), then remove my flannel overshirt and open it on the ground. I cut about a third of the mushrooms away from the bunch, place them in the center of the shirt, collect my tools, tie the shirt off into a makeshift bag, and walk back, certainly looking rather strange to anyone who saw me.

The mushrooms sit on my kitchen counter in a bowl. I am thrilled by them, but intimidated. I decide to do a spore print, something I see many people in the online group recommending. I open a new browser window and type in "How to make a spore print." The first result is from the North American Mycological Association, which recommends slicing the stipe away from the cap, placing the cap face down on a piece of aluminum foil, and covering it with a cup or a bowl for between two and twenty-four hours.

I follow their instructions, forcing myself not to peek for at least two hours.

At two hours on the dot, I anxiously lift up the cap, revealing a perfect imprint of the gills made of powdery, snow-white spores.

I am enraptured. This is pure delight, something entirely new.

A whole new world—one that has always existed but has been hidden until now—is coming to life before my eyes.

I'm going to be obsessed with this, I realize.

This.

Is.

Incredible.

I join every mushroom page I can find from Pennsylvania to Illinois, poring over hundreds of photos and voyeuristically lurking in the shadows, following the comments section for each identification request as they start to roll in, my own version of reality TV. I never dreamed there could be so many different species. It is as though I have lived the whole of my life in darkness and have only now begun to see. Mushrooms are everywhere, and for the first time, they all have their own features; some are exaggerated like cartoons and others are indistinct until I get up close.

One day, with my eyes newly opened it seems, I spot something

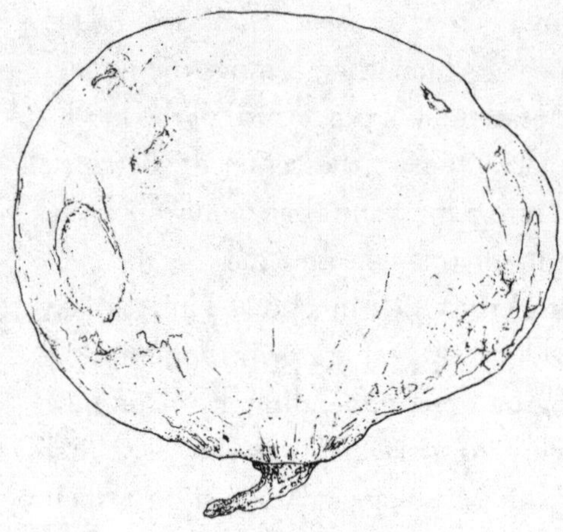

***Calvatia gigantea*, Giant Puffball**

impossible along the fence line of a local park—a perfectly spherical object twice the size of a volleyball, crisp and white with a velvety-soft surface, looking so out of place that it must have fallen from outer space. I look around, wondering if someone has lost some sporting equipment, but nobody seems eager to claim it. I don't even think anyone has noticed it. When I attempt to lift it up, I realize it is attached to the soil.

A mushroom?

My heart skips—I've seen this in photos in the groups, but nothing could have prepared me for the sheer *scale* of it.

A giant puffball!

Giant puffballs are typically classified in the genus *Calvatia,* in this case, *Calvatia gigantea.* There are dozens of different kinds of puffballs, called "gasteroid" fungi for their fascinating reproductive method of developing a huge ball of spores inside a fruiting body, then releasing them in dusty *poofs* after the mushroom has matured. The spores are of poor quality, hydrophobic with a low rate of germination, but the puffball makes up for it by producing them by the trillions.

I've found hundreds more since that day, but as with many meaningful moments, you never forget your first, and it still fills me with delight every time I find another.

Puffballs are among some of the easiest mushroom shapes for beginners to learn on their identification journey, as it is one of the few fungi that adheres to something that could be called a "rule" of edibility. *Edible* puffballs will, without exception, be rounded or lumpy with no stipe, a pure white cross-section that squeaks like Styrofoam when your knife cuts into it, and no hint of other familiar mushroom parts on the interior (for example, gills, which may indicate the immature form of another type of mushroom). Small ones like *Apioperdon pyriforme,* the pear-shaped puffball, will grow on wood, while others, like *Lycoperdon perlatum*, the studded puffball, will grow out of the ground.

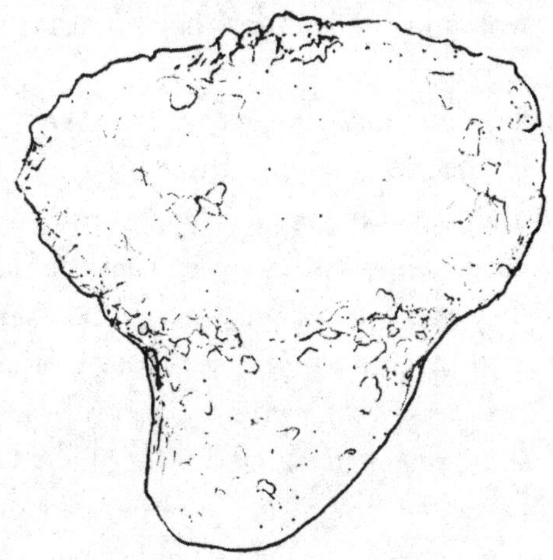

Apioperdon pyriforme, **Pear-Shaped Puffball**

If you are feeling overwhelmed by all the scientific names, you are not alone. Many of these names, called *Linnaean binomials* after the scientist Linnaeus, who created our modern system of taxonomy, stem from Greek or Latin, but learning what a scientific name *means* can help you remember what features make that mushroom unique. Once you start breaking the names down, you might be surprised by how much you didn't realize you already knew.

Take the word *Lycoperdon,* for example. The word *lyco* is lifted from Latin and means "wolf." If you read a lot of fantasy novels, you might have come across the word *lycanthrope,* which some authors use as a fancy term for werewolves. The second half of the word, *perdon,* is a Latin word that I promise you will *never* forget—it means "to pass gas." You might have even misread the word for a modern English word that we use all the time—the word *pardon,* as in "excuse me!" Therefore, *Lycoperdon* means "wolf farts!" (See? I *told* you mycologists were silly people!) What happens when you squeeze an older puffball? It releases spores in a puff of smoke, almost exactly like a cartoonish fart. A newer classification, *Apioperdon,* retains the *fart* part of the name and changes the *wolf* to the Latin word *apio,* meaning "celery." (I know, I don't get it either.) Regardless, when you know a little bit about what the words mean, you can attach those words to the features they describe, which makes them a lot easier to remember.

Sometimes, the meanings aren't so obvious, sometimes they're highbrow, and other times they're just plain funny, the brainchildren of geeky scientists with a comedic streak. Many organisms have been named using words that are made to *sound* Latin even when they aren't *actually* Latin. A good example of this is a fungus discovered by a group of researchers in the genus *Spongiforma* (when we see the suffix *forma* or *formis* we can usually substitute it with *-like*, as in *sponge-like*). Under the microscope, they discovered that this particular fungus had a rather pronounced square shape. Naturally, when they described it, only one name would suffice: *Spongiforma squarepantsii.* Who lives in a pineapple under the sea? This fungus, apparently.

Names matter, but they aren't everything. In my daily life as a freelance composer and academic, I spend quite a lot of time considering the means and mechanisms by which humans create and categorize sound, and I am particularly interested in the way that humans interpret (and often misinterpret) the communications of nature. We may listen to birdsong and perceive simple beauty, yet the bird is communicating—possibly warning of danger, looking for a mate, or attempting to intimidate a rival. We may see a forest full of dense trees and perceive the serene beauty of nature, but within the soil are crowded environments filled with billions of insects, bacteria, yeast, fungi, mammals, plants, eukaryotes, and other creatures communicating through song and pheromones as they entangle, battle, and compromise. Many of these communications go completely unnoticed by us, for we have far bigger things to listen for, but too often we quite literally miss the forest for the trees, as they sing choruses that our ears are too limited to hear.

Honey mushrooms *Armillaria* and *Desarmillaria* are such organisms. They reproduce clonally, aggressively, and to devastating effect. Beginning as a single spore, honey mushrooms grow rapidly, their strong rhizomorphs like blackened fingers clawing deep into trees and leaving the exposed flesh vulnerable to insect predation. Insatiable saprotrophs, the fungi make quick work of the dead wood, extracting its sugars and producing huge fungal blooms, which waft spores off onto the breeze to continue the honey mushroom's sylvan conquest.

They are neither cruel nor kind.

They are, like us, simply *hungry*.

THE HONEY MUSHROOM embodies the settler's dilemma. It is in the mushroom's nature to consume. Fungi stake their claim on trees and in the soil, much like how human colonizers have declared their territories on scraps of paper made from the same trees. The honey mushroom is a sentinel, bearing witness to human conquest, yet offering a lesson for those who take the time to hear it.

Winter is coming for fungus and man alike, and no one understands that better than the honey mushroom. Because the honey mushroom is *more* than just a mushroom, beyond the individual. It belongs to a fungus, which belongs to a tree, which belongs to the forest, which belongs to all living things. A cycle unbalanced will spin itself out into oblivion. A fungus with nothing to eat and nowhere to hide will starve. A system that does not operate in mutualism, in agreements, in respect, will eventually go dark.

It is nothing to know the honey mushroom's name if you shut your ears to its lessons.

It is nothing to own the whole world if you choose not to belong to it.

AUGUST 2022

Allegan, Michigan

The ever-shortening days are a welcome change from the oppressive crush of summer heat as my friend Rudy and I crunch our way through the foliaceous blanket of autumn. There is a tenuous softness to the air, an exhalation of sorts—it feels as though the whole of the forest has opened its windows and let a breeze in.

Early autumn is a wonderful time to be a forager, but especially to be a mushroom hunter. As the massive buildup of organic matter created during the hot summer months begins to spend itself upon the forest floor, fungi are given the delightful task of consuming all the excess. When conditions are favorable, this results in an explosion of quickly-appearing fruiting bodies, the joyful creations of satiated fungi, which usually means well-stocked larders for happy foragers of all species.

Rudy and I are taking advantage of this sudden glut of fungal activity. I have my basket, and he has the maps. We jokingly call him "The Analyst," a reference to his knack for scoping out new locations using Google Maps. Once we arrive, I am set loose on the landscape like a dog on a hunt.

As much as I love to experiment in the kitchen, I really don't enjoy cooking for myself. I like having food *around*, I like eating (perhaps a bit too much, and admittedly a bit too often), but what I *really* love is cooking for other people. Making a meal for someone has always felt like the most natural way to love them.

It doesn't seem worth the trouble when it's just me.

Fortunately, having Rudy around makes it worthwhile to cook again. A more enthusiastic taste tester could never be found. From tried-and-true winners to experimental flops, Rudy will try *anything* I give him. Foragers tend to be willing to try just about anything.

"Oh hey!" I exclaim. "Highbush cranberry!"

"Is it edible?" he asks, eagerly picking a bright red berry and examining it.

"Yes—" I begin to respond as he immediately pops it in his mouth.

"—but not like that." I finish. I watch his face, cringing as the acrid flavor hits his taste buds and his face screws into a pained grimace. He'll be spitting that out for a while.

Sometimes his excitement gets the better of him.

He looks for a moment like he might be sick, then shrugs and grins. At least he's a good sport about it. A lesser man might be embarrassed.

This is a good year for mushrooms. Massive, auburn clusters of honey mushrooms (*Desarmillaria caespitosa*, the same species as the first mushrooms I ever foraged, but with a shiny new name, as so often happens in such a young science as mycology) have popped up at the base of nearly every tree. They're early this year. I've gotten to know them better, to understand their peculiarities and grapple with their myriad lessons.

Russula of all colors are scattered beside the path as far as the eye can see, as if they had been dropped there like breadcrumbs for us to spot. I poke around hopefully, searching for lobster mushrooms, *Hypomyces lactifluorum*, the result of a parasitic takeover by a curious ascomycete fungus that transforms a mostly unappetizing *Russula* or *Lactarius* into something extraordinary—a crusty orange mushroom shaped like a lobster's tail with a firm, dense texture. I strike out again, as usual, but there's always next year.

Amanita muscaria var. *guessowii*, the yellow fly agarics, sparkle golden in the sunlight, hues ranging from lemon to tangerine, dotted with chunky white warts and sporting ragged stipes. Squinting my eyes, I can just about make out the shape of a cluster of creamy-gilled oyster mushrooms, too high to reach.

MUSHROOMS NEVER STOP inspiring me. I marvel at how different and how similar they are, both in function and form, but even in their varying ingenuities and personalities, their likes and dislikes. They are so startlingly human at times, or, rather, perhaps humans are just magnificently fungal. Something that has always struck me is how often people express a fear of "toxic lookalike" plants and fungi, citing these brazen liars lurking on the periphery of our cousin taxonomic kingdoms as the main reason why they "could never possibly forage." When I press these timid would-be foragers about *why* they feel this way, I can generally draw out one reliable response (paraphrased below):

"There are too many things out there that will kill me and I can't trust myself to know the difference. With my luck, I'd mix the good thing up with something poisonous and be dead by the time it hit my stomach."

Controversially, perhaps, I have been quite vocal in my belief that a plant or a mushroom will *always* tell us who they are if we bother to spend enough time examining the evidence they have so graciously provided to us. The subtle differences in their morphology are the key to unlocking exactly who a mushroom is, from the striations along the edge of a cap to the swell of a volva hidden under the earth, the details we often overlook can tell us the most. Despite certain unfortunate naming conventions (looking at you, whoever named the amethyst deceiver!), fungi are utterly incapable of pulling a fast one on us. We are simply . . . impatient. To blame a fungus for daring to injure us is to reveal a somewhat misaligned anthropocentricity, as though our human needs are the only ones worth serving. (Not to mention how bizarre it is to shirk one's personal responsibility in matters of identification prior to consumption!)

During a profound moment in Oscar Wilde's *The Importance of Being Earnest*, Algernon remarks that "the truth is rarely pure and never simple." It is a sardonic twist on a cliché, and an unfortunately accurate observation that humans are constantly making concessions and excuses for ourselves. Someone you've known for decades can keep secrets from you that, when found out, can practically cause that person to change species. The actual essence of a human being is far harder to identify than even the most baffling fungus—we happen to be excellent liars, hidden in layers of well-practiced skins that we wear like cloaks. Perhaps this is why we are so reluctant to trust our own brains and eyes.

Amanita muscaria var. *guessowii* is the first fly agaric I ever found. It is more recognizable in its red and white form, ubiquitous across continents and in popular media, but it retains the same air of mystery, almost a spirituality, in any shade. It is a mushroom with many masks—in one sense, it is somewhat toxic, containing ibotenic acid, a neurotoxin, and in another, it is psychoactive, containing muscimol, which can produce bizarre hallucinations and delusions. In some cases, it can be a medicine, having been used by healers for centuries in controlled doses to treat nervous system conditions, and when prepared correctly in two changes of boiling water, it can be a delicious edible, as all the aforementioned chemical compounds are water-soluble.

My fear of human betrayal runs deep in my marrow. I am slow to believe in the goodness of people, suspicious of their motives, anxious over their like or dislike of me.

I, too, have many masks.

RUDY IS A slow walker, endlessly captivated by the microscopic details of lichens and fossils. I move through the forest much faster, predicting what I will encounter based on patterns of color and trees, topography, and the movement of water. We lose each other with some frequency, until one of us calls out excitedly to the other to come look at something.

"Chicken!" I shout back down the path. "Come look at this, it's a perfect *cincinnatus*!"

I hear a faint "Nice!" in response.

This time of year, my basket is liable to be over-full more often than empty. The forests are ripe to overflowing, and every producer in the system is giving one final push, spending every last bit of remaining photosynthesizing power on producing fruit and nuts, stocking up sugars in their roots and tubers. The mushrooms are blooming more vigorously than usual, sensitive to the impending frosts that my own bones cannot yet anticipate.

Laetiporus cincinnatus, or the white-pored chicken of the woods, is a polypore with magnificent texture and mild flavor, and can operate as a practically perfect stand-in for chicken breast in any recipe. When found, it typically grows in a rosette on dead or dying trees, and may be sprouting from buried wood or tree roots. The mushroom will range from bright orange to a soft, sunrise pink, with a cream-to-white underside and squishy, pillowy fronds, particularly when young. It is the best variety of chicken of the woods I have ever tasted, with a texture superior to both *Laetiporus sulphureus* and *L. gilbertsonii*, even when picked a bit past prime.

I sit with quickly waning patience as Rudy takes his time catching up. I learned a while back that he wants to take photos of everything, untouched and perfect, but waiting for him to arrive can border on excruciating.

"Whoa!" I finally hear behind me. "It's beautiful!"

"I know!" I say, beaming. "Not sure I've ever seen a prettier one in my life."

I move out of the way and let him sweep away leaves and sticks to stage the perfect photo. It amuses me that he wants everything to seem entirely untouched by human hands, but that he is so precise with his environmental primping. An absurdity, a contradiction, but a harmless one.

Once he is satisfied, he lets me take over. I brandish my mushroom knife and begin the process of field-butchery, carving off one frond at a time and brushing away debris until I have carved away a satisfactory amount of the mushroom. Each piece of this chicken of the woods gets a thorough examination and any bits of wood, dirt, or hidden bugs get a swift eviction before being placed neatly in the basket.

"What do you think you're going to make?" he asks eagerly. "Remember the tacos?"

Oh, yeah, I think to myself. Those tacos were *good*. At that moment, an idea strikes me.

"Actually, maybe I'll try making something you've never had before."

This clearly pleases him. Something new is even better than an old favorite. "New experiences" is practically Rudy's mantra.

My mind often wanders while I'm foraging. Every edible mushroom and berry and leaf is brimming with culinary potential, and for whatever reason, I feel more adventurous when I'm working with forest food. The wild tends to inspire, to give me space to dream. Wild food seems to carry extra meaning within its context, unlike the carefully curated selection of produce available at my local grocery store.

We continue on, splitting off occasionally, dreaming of mushrooms and savoring the rare air of autumn.

I look forward to hikes with Rudy. This is atypical for me—usually, I don't much like foraging with anyone, but we have a rapport. He seeks, I find. He tolerates my hyperactive inconsistencies and constant need to rush ahead, and I deal with his infuriating moseying pace. There is an openness to him, a generosity of spirit that I find calming to be around. He is a large man, towering over me at six and a half feet tall, but with a voice low and quiet, like a mighty oak standing guard over a halcyon meadow. Normally I am suspicious of large men, but if you met Rudy, you'd understand how ridiculous it would be for anyone to fear him.

He is always trying to make himself smaller, slouching, kneeling, taking up less space.

I suppose you never get used to being unable to hide, always being bigger than everyone else.

It is no surprise that someone like him would be most at home among trees.

MUSHROOMS ABOUND—THEY SEEM to be fruiting everywhere. I collect the last of the cinnabar chanterelles, we inspect endless carpets of moss bejeweled with brightly colored *Amanita* and sweet-spiced corrugated *Lactifluus*. I bound ahead, he catches up. We walk like this for hours.

"This is spicebush, *Lindera benzoin*," I tell him excitedly as we approach a shrub covered in red berries. "See these? They taste like cardamom, allspice. I made cookies with them last year, you would have loved them." I crush a berry in my fingers and pass it to him to smell. He misinterprets the gesture, popping it directly in his mouth.

"I meant to smell it, not eat it!" I exclaim, wheezing with laughter. "You don't eat them fresh!" His eyes widen and he starts spitting furiously.

We both double over, cackling and holding our sides. "That's *twice* today, Rudy! You have to listen to the end of the sentence, man!" I sputter, choking as I gasp on my words.

He recovers, wiping tears of laughter from his eyes. "I'd say I'm embarrassed, but . . . I'm not," he shrugs, shooting me a cheeky grin. "I just love treats."

WE CARRY ON, pointing at mushrooms and trees, eyes peeled for edibles (me) and anything of scientific value (Rudy).

Dusk begins to lengthen our shadows as we approach the parking lot for the car. I pause—how did I miss it?

"You want a red fruit that actually tastes good raw?" I ask.

"Promise?" He narrows his eyes in mock suspicion.

"Pinky swear." I grin. "Look!" I lift a branch from a nearby shrub, revealing dozens of football-shaped fruits the size of my thumb. "Cornelian cherries!" With the unintentional urging of my slight action, a dozen plump fruits fall to the ground, a few landing in my basket.

Cornelian cherries, *Cornus mas*, aren't cherries; they are a species of dogwood native to western Asia and south-central Europe. The fruits aren't named after someone named Cornelius either; their name is actually a reference to a semi-precious stone called *carnelian*, a reddish form of quartz. In some literature, you will see the berries referred to as "Carnelian" cherries, referencing their similarity in color and the oblong, oval shape in which carnelian is often cut for jewelry. The berries have one long, pointy-ended seed in the center, jammy flesh full of pectin, and a range of flavors. Some fruits may remind you of raspberry, others of currants or cranberries. I've tasted some that bear an uncanny resemblance to the flavor of fresh cling peaches.

Rudy immediately pops a berry into his mouth, sucking the flesh away from the seed. I watch a look of surprise flash across his face, almost immediately replaced with excitement. "Let's bring some back!"

I show him how to select the best ones—how to always take the darkest and squishiest berries, especially ones that have recently fallen to the ground. They seem overripe at first, but a taste test declares them utterly perfect. We fill two empty coffee cups in no time, working intently without speaking.

Rudy's voice breaks the silence. "You know," he says, "people say that what doesn't kill you makes you stronger. But it's not true."

"Yeah?"

"Yeah. Whatever doesn't kill you actually just makes you weirder and harder to relate to."

"I guess." I pause and turn to the field, watching the purple bergamot and echinacea flowers bob left and then right in the breeze.

The berries are staining my fingers red, matching the skin around my ratty fingernails. I've chewed them raw—fingernails, cuticles, an ancient callus on the knuckle of my left thumb. I've been pulling my hair out in chunks for years, eventually disguising my growing bald spots with a crisp side shave.

Rudy looks down at my right hand and lifts it up to his eyes, examining it.

"You have beautiful fingers," he remarks, apropos of nothing, then pauses. "Especially when they're full of berries."

WHEN I FORAGE, I often run across plants and mushrooms that I have never seen or taken note of before, and as a curious person (and an insufferable know-it-all), I like solving the puzzle that a new botanical or fungal friend presents. However, there is always the danger that one's excitement may outweigh their critical thinking, especially if they have significant experience on their side. Too often, I have seen trigger-happy foragers fry up what they were positive *had to be* a lovely haul of parasol mushrooms, only to find themselves absolutely wrecked from both ends scant hours later by the similar-looking *Chlorophyllum molybdites*, aptly named "The Vomiter."

And that's the *better* scenario. Worse, you could end up like the

German composer Johann Schobert (not to be confused with the far more famous Franz Schubert), who foraged a great bounty of mushrooms and brought them to a chef to be prepared, only to have the chef refuse and inform him that they were poisonous. Undeterred, he found a second chef, who echoed the warning of the first. Finally, he sought the advice of a physician, who offered the eighteenth-century version of a shrug and an "I dunno, seems fine to me." Vindicated by the doctor's inexpert advice, he took the mushrooms home, made a soup, and promptly killed himself, his wife, all but one of his children, and the doctor friend, who probably wished that he had never accepted Schobert's dinner invitation.

A few lessons can be learned from Schobert's folly. First, bad advice is free and abundant, and it often comes from those operating outside of their scope of practice. Second, it is always wise to heed the warnings of people who work with food for a living. And third, an overly cautious forager may experience an outsized measure of disappointment, but at least being a skeptic lets you live another day.

Again, with the best culinary wisdom of our time, Anthony Bourdain said "good food and good eating are about risk." On the surface, this seems reasonable. When you go to a restaurant, you don't assume that the line cook has been seasoning your fries with arsenic, but even that tiny, calculated measure of trust is necessary in order to convince you to dip that fry in ketchup and take a bite. More trust is required in the case of sushi, where you may choose to put your life in the hands of the itamae in order to taste fugu, a dish that is perfectly safe when made correctly, but which can turn fatal with a single error in knifework.

The forager, unlike a diner at a restaurant, has to possess an unusual degree of confidence. We have nobody but ourselves to blame if things go wrong. But how do you handle assuming all of the risk of your food, from start to finish?

For one, we need to consider food safety and how it is implemented in the commercial space versus the home kitchen. A frequent concern that I address for anxious new foragers has to do with environmental pollution. "How do I know that the area where I'm foraging isn't contaminated?" is probably among the top five questions that I respond to. I find it a helpful thought experiment to ask "How do you know that the area where my grocery store food is grown isn't contaminated?" And in reality, unless you're extremely vigilant, you can generally assume that it is. Herbicides, pesticides, and fungicides are necessary for maintaining pristine, massive monoculture crops, and unsafe levels of heavy metals are common in these environments. In fact, the use of hefty deterrent chemicals that have known harmful effects on humans is one of the main reasons that I feel far less comfortable foraging asparagus on the edge of a farmer's field than I do picking fruit in the city.

Other risk factors to consider in grocery store foods, particularly produce, are those of bacterial contaminants, like listeria. Commonly associated with the consumption of improperly stored deli meats and cheeses, listeria is also a common passenger on produce items like fruit and leafy greens. This and similar contaminants may be introduced at any stage of the food process, from harvest to packaging and handling, to any number of fellow customers who may have pawed over those very same nectarines you've chosen to load into your cart. How many people do you suppose have touched your food, even when it appears to be nestled safe and secure in its plastic packaging? Every time we pick up a piece of fruit, we are placing tremendous trust in dozens of people who may not have the same health and safety standards as we do.

This isn't to say that there *aren't* significant considerations to keep in mind regarding pollution or safety when foraging. We must be constantly vigilant and aware of our surroundings. The ability to recognize signs of spraying in public areas or along nature trails is almost as important as the ability to identify plants and fungi for consumption, because herbicides can render what would normally be identified as a "safe" food quite the opposite. Some of these signs include flattened, dead plants; sick-looking or floppy stalks; odd coloration; a whitish, powdery cast on the surface of a plant that shouldn't have it; or areas of bare, uncovered soil (often found at the bases of trees in public areas).

Sometimes it is a literal sign (often required by law, especially near wetland areas) indicating that the area has been sprayed, and will often be accompanied by a warning to avoid interacting with plants in the environment for a certain period of time. The use of various 'cides on invasive plants is quite common in cities and on nature paths, and may not be immediately visible to the naked eye. Many of us who forage have at one time or another fallen prey to surreptitious toxicity in the form of a recently sprayed stand of knotweed or garlic mustard. Heavy metal toxicity is a very real concern, especially on land that was or is used in ways that increase the amounts of heavy metals in the environment (for example, near highways, in or around contaminated water, or close to mechanic shops or factories).

As far as the ecologic effects, the use of certain 'cides isn't all bad. Kyle Lybarger, Alabama forester and conservationist and founder of the Native Habitat Project, frequently demonstrates how the targeted use of certain chemicals can help to *increase* biodiversity by reducing or eliminating invasive competition, particularly for more aggressive or stubborn invasive plants. Often, the deft use of certain herbicides is the most powerful weapon a conservationist has at their disposal to begin the process of remediating land that has been poorly managed.

Another factor to consider is not simply how many people have been involved in the processing of your food, but *how much each of those people was paid*. When you think about it, the fact that dozens of people are involved in getting a potato from the field to your table should mean the potatoes cost more. That you can buy five pounds of potatoes for under ten dollars is a clear sign that someone along the way (or, more likely, a *lot* of someones) is being exploited.

In systems that rely on exploitation to function, your food will always be stained with something.

We are, like it or not, bound to the past, present, and future of what we eat.

THE PROCESS OF weighing risk is highly personal. I, for example, prefer some risks to others. Questionable leftovers? Sure, I'll bite.

Vulnerability with important people in my life? Hard pass. (*I'm working on it . . .*)

THERE ARE SEVERAL steps you can take to increase your confidence in what you're foraging. For starters, avoid train tracks and busy roads. Train tracks are often heavily polluted with diesel and other unwanted contaminants, but importantly, they are generally unsafe places to forage. If you're not paying attention, you could find yourself in serious danger. Never collect food from dump sites.

As a general rule, it is best to avoid aquatic plants near roads or in high-traffic urban environments. Aquatic plants tend to be bioaccumulators, acting like environmental sponges, sucking up the heavy metals and chemical toxins that are present in contaminated water. Any plant growing in water adjacent to roads or industry can be assumed to be unfit for collection, though such plants do serve an important ecological function through their water filtration abilities. To be safe, avoid areas where those toxins are present and only collect aquatic plants from clean water far away from potential contamination.

In addition, I cannot recommend collecting fungi from treated wood (you wouldn't believe how many oyster mushrooms I've found growing on telephone poles!). It's tempting, but it just isn't worth it.

Keep in mind that the vast majority of toxic substances accumulated by plants will be contained in the root system, and fewer contaminants will be present as you move up the plant. This means that while it might not be smart to eat a chicory root collected from a busy city street, you shouldn't be concerned about picking an apple from the tree right next to it.

There is always risk. Sometimes it is inherent, and other times it is manufactured.

It is our job to perform the calculus of risk. At what point is a risk too costly if it fails, and who gets to decide?

To save both your gastrointestinal tract and plumbing bill some serious trouble, I highly recommend trying out a system I like to teach all of my classes called the "Three Times Method." (Gosh, I need a better name for that!)

With the Three Times Method, the goal is not so much dinner *today* as it is dinner *eventually*, and even more than that, finding and observing the organism in its natural environment *prior to any practical human use.*

When you find a new plant or mushroom, the first thing you should do is methodically observe each feature and try to match it to a description in your field guide (you can use what's called a dichotomous key for this). Note the environment you found it in. Are there trees nearby? Is the soil loamy, rocky, sandy? What time of year is it? Can you note any specific smells? If it is a mushroom, what happens when you cut it in half, and if you nibble off a piece of it, what does it taste like? (I'm being serious—your mouth and nose are two of the best tools a mushroom hunter has.) Draw the organism and be as detailed as you can. It doesn't matter if you're a good artist or not, because the point isn't making a great piece of art, it is about spending significant time studying your new plant or mushroom friend. Don't collect this one.

Next, find another example of your plant or mushroom in a different location. Take note of the environment. Are there other specimens nearby? Do all the characteristics of this one match the first one you found? If there are enough other specimens nearby and the organism is not otherwise threatened or endangered, you may collect it and bring it home. Are there changes in color or scent when it dries out?

Finally, locate a third example and repeat all these steps. If you are, as Sam would say, "banana-confident" in your identification, try bringing it home and preparing it, being mindful that there may be specific safety preparation recommendations, then eat a couple bites and wait a few hours to test for any personal negative reactions.

Freedom and risk are a mixed drink, no more separable than the forces holding our cells together. They are a package deal.

The trick is learning to make the ratio palatable.

And some people are worth trusting.

When Sam bounds through the woods to show me a particularly special plant he has found, I follow. When he introduces me to that plant, tells me her name and her properties, I pluck off a leaf and munch without hesitation. I trust Sam implicitly, the kind of trust that would have me leaping off a cliff if he told me I would be safe (especially if he promised that there was some obscure sea vegetable waiting for me at the bottom). When my friends Betsy or Alexis hand me a jar of some unidentifiable wildcrafted preserve, I don't question their expertise, nor am I suspicious of any nefarious intentions. I just gratefully spread it on my toast (and it's *always* delicious).

Being able to trust them has its very sweet rewards. But it's because I know they've worked hard for the wisdom of their finds.

I have often been on the receiving end of bad advice offered by the underqualified, with safety "tips" like "cook a mushroom with an onion and if it turns black, it's toxic, but otherwise it's edible," "you can eat any mushroom with a peelable cap," and even "all toxic mushrooms are brightly colored, everything else is safe." (To be clear, all of these are falsehoods.) I have also witnessed lazy identifications, misleading (or just plain inaccurate) literature, and the lofty claims of fools who call themselves teachers, those who lie about their own expertise to gain praise, attention, or followers. I once attended a class led by a "plant expert" who, when asked directly about a plant he had not planned on discussing, fabricated multiple identifications rather than admit that he simply didn't know what it was.

Some people aren't worth trusting.

THE HARDEST OF all is to learn which one *we* are. Am *I* worth trusting? Can I be humble, can I be confident, can I ask questions, can I answer them?

What am I willing to risk? What is it worth to me? More importantly, *who* am I willing to risk? What are *they* worth to me?

Without taking risks, we would never dive for pearls.

Europeans would have never eaten tomatoes.

None of us would ride in airplanes.

Dogs would still be wolves.

There is risk in being human, as there is such a very great deal out there in the cosmos for us to fear. But can there even be this gorgeous feeling of teetering into the unknown if not for the stupendous and terrible moment where we are careening through space with nothing solid beneath our feet and no sense of direction?

The incalculable risk, the foolishness, the disgusting wonder and beauty of it all; could it be that these are what it means to be human?

Whether we choose to venture out or not, we cannot escape it. It is in us, and we carry it wherever we choose to go or hide. It *is* us.

There's no point in running, and there's no point in being scared.

Try it. Maybe you'll like it.

GATHERING EXERCISE

"TASTE"

Go find a food that you have heard someone claim "tastes bad."

It might be a strong-tasting food beloved by a culture unlike your own, such as anchovies, natto, Limburger cheese, or kimchi.

It might be a wild food that you know is not toxic but is deemed by some to be unpalatable, not worthwhile, or unpleasant, such as crabapples, burdock, or slippery *Suillus* mushrooms.

Imagine, for a moment, the opposite of what you have heard about this food.

Make your head open and empty, put the food in your mouth, and enjoy trying it, even if you do not like the way it tastes.

Fried Oyster Mushroom Sandwich with Smashed Blackberries

Serves 4 to 6

It is *very* unusual for me to recommend battering and frying a mushroom, but in the case of oyster mushrooms, even I must make an exception. Every type and species of oyster mushroom is *lovely* when battered and fried, with a crispy texture and a mild, almost sweet flavor that could turn even the most hardened mushroom hater into a lifelong mycophagist. They don't taste like oysters—more like a fantastic fried chicken sandwich. I like tucking them into burger buns with smashed blackberries and some dandelion greens. The acidic sweetness of the blackberries softens the bitterness of the dandelion greens, and the distinct textural layers make for a surprising yet delicious bite.

Pro tip: When you're collecting your dandelion greens, only collect half of the leaf (the end closer to the tip), and try to gather from shady areas. They'll be far less bitter. You can also "force" dandelion greens in your yard by placing a bucket over them to shield them from the light for milder flavor and a more tender texture.

4 eggs
1 cup all-purpose flour
¼ teaspoon red garlic powder
¼ teaspoon onion powder
¼ teaspoon smoked paprika
1 teaspoon kosher salt
½ teaspoon cracked black pepper
1 cup panko breadcrumbs
2 cups whole oyster mushrooms
2 cups vegetable or canola oil
½ pint fresh blackberries
1 tablespoon lemon juice
4 to 6 hamburger buns
1 cup dandelion greens

(continued)

Beat the eggs in a shallow bowl at least as wide as your largest mushroom, then set aside. Sift the flour, garlic powder, onion powder, paprika, salt, and pepper into a separate, similar-sized bowl and set aside. Finally, put the panko in a third bowl.

Line up the three bowls in your workspace (first the panko, then the wet mixture, then the dry mixture). Place a tray nearby to hold your mushrooms once you have battered them.

Dunk each mushroom into the wet mixture, then the dry mixture, then the wet mixture again, and finally the panko, then put them on a cookie tray.

Heat the vegetable oil in a Dutch oven or deep-fryer to about 350 degrees or until a droplet of water sizzles, then fry your breaded mushrooms in small batches until each is golden brown with a rich, fragrant scent. Place each fried mushroom on a kitchen cloth or paper towel to absorb any extra oil.

Using a small bowl and a spoon or a mortar and pestle, crush the blackberries until they form a rough paste, then mix in the lemon juice.

Toast your buns, if desired. To assemble the sandwiches, place a small handful of dandelion greens on the bottom bun, followed by a few fried oyster mushrooms on top of the greens, then spread a generous layer of blackberry paste on the top bun. Serve with roasted root vegetables, potato chips, or a side salad with even more blackberries and a vinaigrette.

Giant Puffball Pizza

Makes 1 puffball pizza

Ah, the pizza. This pizza was one of my most viral cooking videos, and for good reason: it's a wacky process with one of the most bizarre mushrooms you could possibly cook. Puffball pizza is easy to make, but it's also easy to make *incorrectly*. The general idea is that a slice of puffball serves as the "crust," and sauce, cheese, and toppings are placed on top. Unfortunately, I know from experience that if you don't prepare your puffball properly, you'll end up with a soggy mess. It may *look* okay, but just wait until you try to bite into it. *Gross.*

This recipe will show you exactly how to make a puffball pizza the *right* way. And remember, if your puffball is any color other than white inside, it's no good for eating anymore. Even if it's *mostly* white, don't bargain or try to rationalize it, don't cut the "bad part" off and try to use the rest, just throw it into the backyard and let nature have it back. At best, it'll taste like feet, and at worst it might even make you sick.

Because of the variable size of giant puffballs and the inherent customizability of pizza, it's not exactly useful to provide precise quantities for the ingredients, but here's everything you'll need:

A pizza stone
A large cast-iron pan
About 3 tablespoons salt
1 giant puffball slice (at least 2 inches thick)
Oil with a high smoke point (refined sunflower oil is a good choice)
A smaller cast-iron pan (one that can fit inside the large pan), or similar flat-bottomed heavy object
Pizza sauce
Shredded cheese
Pizza toppings of your choice

Preheat the oven to 450°F with the pizza stone inside.

On the stovetop, heat your large cast iron up and add a generous amount of salt to the bottom. When the cast iron is hot, add the slice of puffball and place the smaller cast iron on top of it, pressing down with your weight. The goal here is to coax as much water out of the puffball as possible (no easy task, since this mushroom is at least 90 percent water).

Check the bottom of the puffball slice occasionally, making sure it isn't burning. When the slice starts to stick, lift it off, add more salt, flip it over, and replace the weight on top of it. Eventually, the mushroom should reduce dramatically in size and may have some slight browning.

(continued)

Remove the puffball slice and add 1 tablespoon of oil to the pan. Finish searing the mushroom by frying both sides evenly in the oil.

Remove from the heat. Transfer immediately (and carefully!) to the preheated pizza stone in the oven and bake for 10 minutes to dry the surface of the mushroom. After 10 minutes, remove from the oven and dress the mushroom slice as you would a pizza, adding sauce, cheese, and toppings, then place it back in the oven and bake until the cheese is bubbly, slightly browned, and melty. Remove from the oven, cool, then slice, plate, and share with someone who thinks they hate mushrooms.

"The Rice"

Serves 6 to 8

My family does not celebrate a single holiday without "The Rice." Birthdays, weddings, Christmas, The Rice is always present, a small but important way that we acknowledge and celebrate our Puerto Rican heritage.

Several years ago, I decided that I wanted to take some steps toward decolonizing my diet—honoring and reclaiming the foods I grew up with by adapting them to fit within the context of my current terroir and the history of my ancestors. The original spirit of The Rice, after all, was ingenuity in the face of oppression—my predecessors managed to work under the crushing weight of colonial rule by creatively blending staple foods from the island with the new foods that were introduced to them. In the same way, I wondered if I could adapt The Rice to the foods I could find in my North American landscape instead of being beholden to Goya imports from a dusty, forgotten section of the grocery store. While a lot of the ingredients in this recipe will seem familiar to my Puerto Rican brethren, a few others will stand out (including the inclusion of *Monarda fistulosa,* also called bee balm or wild oregano).

This is the recipe where you'll want to use the Wildcrafted Sofrito you made earlier in the year (page 71). If you didn't make it, no worries, you can buy it in the frozen section (or make your own seasonal version), but keep in mind that you'll probably need to double the amount if you use store-bought. It has occurred to me that some people might see my inclusion of a mushroom that is not typically eaten in Puerto Rican cuisine as some sort of American blasphemy, but to those people, I will gently suggest that they search "*Laetiporus*" using iNaturalist's explore function and see what shows up over the island.

¼ cup vegetable oil
1 yellow onion, diced
3 small garlic cloves, minced
¼ cup sofrito, homemade (page 71) or store-bought (double if using store-bought)
1 teaspoon ground coriander
1 teaspoon ground cumin
1 tablespoon ground annatto seeds (achiote)
1 teaspoon garlic powder
1 teaspoon freshly ground black pepper
1 teaspoon fresh bee balm leaves (substitute oregano if needed)
1½ cups diced fresh chicken of the woods (½-inch cubes)
1 (7-ounce) jar alcaparrado (green olives and capers)
3 cups extra-long-grain white rice
1 quart chicken or vegetable broth, plus more if needed
2 (15-ounce) cans gandules (pigeon peas)

In a caldero or Dutch oven, heat the vegetable oil over the stovetop until it is hot enough that you can add a drop of water and hear it sizzle. Next, add the onion, garlic, sofrito, coriander, cumin, annatto, garlic powder, black pepper, and bee balm, stir to coat, and sauté until fragrant and beginning to deepen in color, about 3 minutes. Add the chicken of the woods and sauté for 1 minute, then add the alcaparrado with 1 to 3 tablespoons of the brine. Once you see everything start to bubble, add the rice and stir to coat. Toast the rice in the oil for 1 to 2 minutes over medium-high heat, stirring occasionally to avoid burning. The oil and the rice will be fragrant and begin to darken slightly in color.

(continued)

Pour the broth into the pot. The rice should be just barely submerged beneath the liquid—if it is not, add more broth or top off with water. Drain both cans of beans, discarding the liquid, and add them to the pot, but do not stir. Cover the pot with aluminum foil, then place the lid on top and let it simmer for at least 20 minutes. Resist the urge to check or to stir—stirring will leave you with mushy rice and bruised beans.

If your rice is nearing doneness but seems wet, remove the lid and foil to help some of the liquid evaporate. If most of the liquid has evaporated but the rice is still crunchy, try lowering the heat and adding a little more broth before you replace the foil and lid.

After the rice is done, remove the lid and foil and let any residual steam evaporate before serving. The best part is the pegao, the crunchy rice that sticks to the bottom of the pot. If you're quick, you might get to it before the rest of the familia knows it's done! ¡Buen provecho!

Cornelian Cherry Fruit Leather

Serves 10 to 20

Cornelian cherry has quickly become one of my favorite fruits to work with in late summer. They're not *actually* cherries; in fact, they are the fruit from a type of dogwood called *Cornus mas*. While the dark red, oblong fruits may bear a passing visual resemblance to cherries from a distance, most of them don't taste a thing like a cherry. Flavors of individual specimens may vary from the likes of cranberry, raspberry, even peach. They make a great trail snack, but due to their high pectin content, they also make superb jams and fruit leathers (if you can put up with de-seeding them, as their odd, double-pointed seeds can be a challenge to work with).

I typically de-seed my Cornelian cherries with an ingenious little device called a food mill. It fits on top of a bowl or a pot and is excellent for separating seeds and skins from fruits and vegetables, leaving you with a fairly uniform pulp. It isn't perfect and the process can leave a good bit of fruit behind, but it does work well, as long as the fruits are *very* ripe. Sometimes, in order to get that last bit of flavor off the seed, I boil the seeds in just a little bit of water and then strain out the juice. By far the easiest way to extract the juice by itself is with a steam juicer. The juice is admittedly a great thing to have around, especially for adding to drinks and making jelly, but if you want fruit leather, you'll need the pulp. Here's how I make mine.

1 pound Cornelian cherry pulp
2 to 4 tablespoons maple syrup
½ teaspoon citric acid

Blend the Cornelian cherry pulp in a food processor or a blender, adding maple syrup just a little bit at a time until you have a thick paste (it should not be runny). When the consistency is smooth and similar to tomato paste, add the citric acid and blend again. Spread the paste in a single layer (about ½ inch thick) onto a piece of parchment paper on a silicone dehydrator sheet, or on parchment placed directly on a wire dehydrator rack. You may be tempted to make your sheets thinner than this, but remember that they will shrink considerably during the dehydration process, and thinner sheets may result in cracking. Dehydrate at 135°F for 6 to 8 hours, or until dry but flexible.

When the fruit leather is done, cut the leather and the parchment into strips and roll up together. I like to tuck them into my lunch bag for a little midday sweet treat, and I also use them when I need a fancy little garnish for a cake or pie. Store at room temperature in glass jars with a desiccant packet for up to 6 months or use vacuum bags to double the shelf life.

Spicebush Shortbread

Makes about 16 cookies

Spicebush is one of the easiest ingredients for new foragers to identify and begin to work with, since it has a lot of visual and olfactory cues that confirm its identity. It's shrubby and spindly, with bright, oblong green and red berries, alternating leaves, little pimply lenticels on the smooth bark, and smelling wonderfully of rich, spiced chai when virtually any part of the plant is crushed, but especially the berries and twigs. Gastronomically speaking, spicebush berries are passingly similar to many familiar spices, like cardamom, allspice, and black pepper, and can be used in similar ways, but at the same time, spicebush is somehow different from all of them. A good way to test the range of this wonderful berry is to make cookies. Resist the urge to add your usual baking spices—I promise you, the spicebush doesn't need cinnamon or nutmeg's help to shine.

2½ cups all-purpose flour
¼ cup cornstarch
1 cup (2 sticks) unsalted butter, softened
⅔ cup granulated sugar, plus 3 tablespoons for finishing (try using your rose sugar, page 118, from earlier in the year for the finishing touch!)
⅛ cup mugolio syrup (page 115) or molasses
1 tablespoon vanilla extract
1 teaspoon salt
1 tablespoon dried red spicebush berries, finely ground (don't be concerned if this comes out a bit pasty, since spicebush rarely dry completely through)

Preheat the oven to 325°F and line a cookie sheet with parchment paper.

Sift the flour and cornstarch into a medium bowl. Using a stand or handheld mixer, cream the butter, ⅔ cup of the sugar, and the mugolio syrup, then add the vanilla, salt, and spicebush. A little at a time, fold the dry ingredients into the wet until the dough comes together. Scoop the dough out onto a sheet of plastic wrap, cover, and chill for at least 2 hours.

Remove the dough from the fridge and roll out to desired thickness. Cut out shapes with cookie cutters or a knife, then bake for 20 to 30 minutes (depending on the size and thickness of your cookies), until golden brown and buttery on the edges, smelling of holiday cheer.

Sprinkle the cookies with the remaining 3 tablespoons sugar and serve warm with coffee or tea.

Honeyed Honeys

Makes 1 pint

I love collecting cookbooks from all over the world, especially old ones. I don't always understand the languages they're written in, but half the fun is trying to figure out the translations. A few years back, I acquired a stack of cookbooks from Eastern Europe written near the turn of the twentieth century, and I was eager to see what sorts of novel recipes I might find for mushrooms. One of those recipes was for what appeared to be honey-preserved mushrooms.

The idea was fascinating—honey is a fabulous preservative, so it could easily work. I took a few fresh porcini, sliced them thin, and popped them in a jar with honey. To my immense disappointment, the mushrooms rotted within days, creating a slimy, disgusting mess and wasting my precious honey and porcini. Immediately, I realized what the problem was: in my clunky translation process, I had somehow missed that I needed to use *dried* mushrooms, not fresh!

This recipe is a play on words, since it uses dried honey mushrooms, though you can substitute with other dried mushrooms you already have in your kitchen from earlier in the season. A few cautions: you can use a *lot* of different kinds of mushrooms with this method, but avoid using any that have specific cooking safety requirements (like morels or chicken of the woods, which may even be considered toxic when raw). It is also possible to make psilocybin-infused honey by using dried "magic

(continued)

mushrooms" depending on the legality of psilocybin possession in your area, but you didn't hear that from me.

I recommend using a heat-based method to dehydrate any mushrooms before infusing them in honey (at least 135°F). That way, you're working with mushrooms that have effectively been cooked, reducing your chances of gastric upset. The honey and the mushrooms have a lovely effect on one another, the mushrooms infusing the honey and the honey soaking into the mushrooms.

About 4 ounces honey mushrooms, thinly sliced and dried
About ½ pound raw honey

Place the dried mushrooms in a sterilized pint jar, then pour the honey over the top. Leave about 1 inch of headspace before screwing on the lid. Store in a cool (but not cold), shaded place. Once per week, flip or shake the jar to prevent the mushrooms from getting stuck above the level of the honey, which may cause spoilage. The honey-preserved mushrooms will be shelf stable for several years, and are best enjoyed after at least 6 weeks. Enjoy as part of a charcuterie board, drizzled on toast with soft cheese, or poured over ice cream.

The Giving Trees (Katajjaq)

You know me, I think there ought to
be a big old tree right there.
And let's give him a friend.
Everybody needs a friend.

Bob Ross

It is impossible to foster any significant love for mushrooms without also developing a great admiration for their partners, the trees.

Trees matter far, far more than you might think when you are hunting for mushrooms. "Sure," you might say, "look for dead trees, you'll probably find some mushrooms on them." And, of course, this is true, but it isn't the whole story. Not by a longshot. For many species of fungi, finding a specific tree is the key to locating a specific mushroom. They are symbiotes, sometimes only present where one or the other can be found. Many fungi, in fact, are named by scientists in tribute to their tree partners. *Hypsizygus ulmarius*, the elm oyster, is just one example. The genus name *Hypsizygus* comes from two Latin words meaning "high up" and "yoked," referring to its habit of producing deeply attached mushrooms high up on the trunk of its host, and the species name *ulmarius* refers to the elm tree on which it can be found. Another example is the scaber-stalked bolete *Leccinum quercophilum*; the species name referring to its "oak-loving" proclivities. Poke around under enough oaks in the right place at the right time, and you'll find it.

***Hypsizygus ulmarius*, Elm Oyster**

A few years ago, I went morel hunting with my friend Glenn, who is, admittedly, a very odd man. Glenn is in his fifties, about five feet three, a vegan, and a musician who runs a recording studio. He hunts mushrooms on the weekends in Grand Rapids, Michigan, sometimes selling his extras, usually trading them by the pound for good meals at nice restaurants. Glenn is the type of person who giggles whenever he finds a morel, unable to contain his joy.

I like hunting mushrooms with Glenn, even though I always get skunked by him.

Glenn always finds morels before I do. Even when I sit in exactly the right spot with my face inches from the ground, scanning the landscape like a satellite, he always finds something I missed. Sometimes it's right in front of me, like a neon sign.

When I expressed my frustration to Glenn about how much faster he could find morels than me, he just laughed at me. "You have to think like a mushroom," he said. "You're only looking on one side of the road, in the spots that make sense to *you*. You're missing all the best ones because you're thinking like a human. Mushrooms travel different roads. Keep looking and you'll see their paths."

It was Glenn's gentle way of saying "Think outside the box, dummy."

If you cannot distinguish a pine from a maple, every mushroom you find will be nothing more than delightsome happenstance. But we can do better; we don't have to bank on accidents.

We can think like mushrooms.

Thinking like a mushroom is no small task. We have to consider what sorts of conditions a specific mushroom prefers, who their likely friends are, collect all sorts of demographic information and use it to determine where they might like to be.

Giant puffballs, *Calvatia gigantea*, for example, like to showboat. They prefer to pop up in places where they don't have a lot of competing organisms who might steal their thunder. Well-trampled places, open forests, pastures, meadow edges, cemeteries, anywhere where the trees (if there are any) are large and spaced far apart, spots where animals and people are roaming around. They like to be seen, and they *will* stand out.

Oyster mushrooms, *Pleurotus* species, are cheerful, consistent, and generally agreeable, willing to grow on practically any type of substrate. From fallen logs to wood chips, trash bags to telephone poles, oyster mushrooms have few qualms about where a meal comes from, as long as it can be consumed.

Black trumpets, *Craterellus* species, love thick moss and oaks, thriving in the moist and shade. They are shy and difficult to spot when compared to their showier cousins. Trumpets follow the flow of water, or sometimes the roots of trees, and they frequent old cemeteries, in keeping with their gothic nomenclature.

It is one thing to know a plant or a mushroom, to be familiar with its chemical makeup and its physical features, its human uses or cautions, whether it is medicinal, edible, or toxic. It is quite another to develop the intuitive sense and observational skill to learn *who* that organism is.

Western science, for all its revolutionary wonders, can be a depersonalizing way of knowing a thing. We observe and describe as neutrally as possible, rejecting anything that walks, talks, or smells like anthropomorphism. This has its uses—we tend to try to relate the world around us to ourselves, assigning human traits and motivations to decisively nonhuman entities, which can be destructive, causing us to misunderstand their behavior. If we try to relate a cat's behaviors too closely to those of a human, we run the risk of ignoring the unique way in which the cat perceives the world and the instincts they possess, which can lead to us disregarding that animal's needs. A cat is a cat, not a person, and that is just as good. Understanding that my cats are fundamentally Not Me allows me to recognize the way that their behaviors are communicating *to* me. Flattened ears tell me "I'm not a fan of you trimming my nails," wide eyes with dilated pupils and forward-facing whiskers say "I *really* want you to throw that toy again," a tail held upright and a high-pitched trill as my cats trot toward me is their way of saying "Hey, it's great to see you! When are you going to feed me?"

Science is pretty good at allowing us to identify the relationships between organisms that have obvious similarities, like myself and my cats. We both vocalize, we both experience the world through the same basic senses, we both have similar bodily functions, overlapping consumption needs, and experience feelings like fear, pain, love, hunger, and curiosity. Traditionally, science categorizes human and nonhuman organisms by their relative relationships to one another. Scientific systems like Linnaean taxonomy (classifying living things by functional and genetic similarity) place biological organisms in categories, first by separating animals, plants, fungi, and other types of beings into their own categories, then further dividing them into smaller and smaller boxes along the way. My cats are related to me, inasmuch as we are both animals, and we are both mammals. Beyond that, we branch off, occupying different subcategories as we go further down the phylogenetic tree. My cats are not humans, they are cats. But the fact that they aren't humans doesn't mean that they are somehow *inferior* to humans, nor does it mean they are not Persons. They are cats, and as far as I'm concerned, being a cat person is just as good as being a human person.

Humans are, of course, also related to non-animal organisms like fungi and plants. We all evolved from the same single-celled recipe, after all, branching off and experimenting with our genes over millions of years to fill out the earth with staggering biodiversity. It can be harder to conceptualize the genetic relationship between a human child and a black trumpet mushroom, but it undeniably exists: humans share anywhere from 30 to 50 percent of our DNA with members of the fungal kingdom. We are not the same, but we are made from the same stuff. What if we looked at the black trumpet mushroom and saw a relative, similar to how we might see a dog or a cat as a part of our family?

Science is a way of paying attention that can provide one with many facts, facts that offer many practical applications, but unless you use it to extend your ability to demonstrate empathy and interspecies relationship, it will be meaningless, mere trivia. A plant or a mushroom is more than just the sum of its parts; it is a whole world, a spirit, breath. The "otherness" of a tree does not make it less of a Person to me, even though that Person is not human. We are related, but not the same. The tree is *different, not less*. It is not my subordinate; if anything, it is my elder.

I have employed a great deal of Western science-speak in this book, through the systematic introduction of the language of plant and fungal anatomy and taxonomy that has become commonplace among scientists and foragers alike. Having a bit of Latin in your back pocket is rather useful, and can lead to a deeper understanding of the observable relationships between certain groups of organisms, but it doesn't tell the whole story.

Trees are keepers of the ancient past and sentinels of the future, experiencing time in a way that no human could ever fully comprehend, and yet, many trees are one of the best indicators that humans have for the passage of time.

The more we learn about plants and fungi, the more we can recognize the eerily familiar marks of intelligence and self-knowledge in their behaviors. Delicate though the line may be between anthropomorphism and personification, we must acknowledge that the personhood of a mushroom is not necessarily like that of a human person. Anthropomorphism assumes that we can only relate to humans, and that to understand a tree or a mushroom, it must behave or look like a person, wearing one of our faces, branches like hands, fruit like ornamentation. It must make decisions the way we do, exhibit the same moral code as us, have the same desires, quarrels, and motivations that we do. To anthropomorphize is to do a disservice to the organism, even when we think we are doing it a favor by "elevating" it to our status as human.

I like to think of it this way: a mushroom is a mushroom person; a human is a human person. A tree is a tree person. No person is better than the other, and in many ways, they may not even be terribly alike. But all of us are *valuable*, equally deserving of deference and respect. All of us have meaning, purpose, wisdom. Each of us has gifts to offer to the world, to one another.

All sorts of people belong here, in the system we share together. Plant people, mushroom people, tree people, animal people, perhaps even human people, when we manage to communicate with one another.

THE PAWPAW PEOPLE, known by their Latin name *Asimina triloba*, give me annual gifts in the form of pawpaw fruit. September is "Pawpaw Christmas" in my house. Their fruit, technically a berry, the largest native to North America, is sweet, slightly acidic, and creamy, like custard, mango, banana, and pineapple all had a delicious, lumpy, very weird baby. Refreshing and soft, they are best eaten as given, freshly fallen, fresh from the forest floor. You never want to pluck a pawpaw from its tree too early, or it won't ripen properly. You must have patience and only take what is given, freely, without demanding, cajoling, or force.

Pawpaw do not *have* to produce fruit for effective reproduction. They can reproduce asexually, putting up new shoots from their root system and creating vast, genetically identical thickets. However, pawpaw are self-infertile, meaning that the male flowers on one plant are unable to pollinate female flowers on the same plant. They *must* receive pollen from a different plant in order to introduce genetic diversity into the patch. Without genetic diversity, the patch, no matter how large, is doomed to be sterile. Unless it has a partner, it will continue to reproduce only by root sucker, consuming the understory of the forest without giving back any fruit.

We can survive alone, but we can only thrive together.

THIS TIME OF year, everything feels like a gift as I gather up baskets full of walnuts, hickory nuts, acorns, and chestnuts. There are times when it feels like the trees are practically raining bounty upon my head (be especially aware of what's happening above you when you're gathering walnuts unless you want to understand this very literally), and when I'm gathering, it almost feels like I'm getting away with something I shouldn't. The pickings

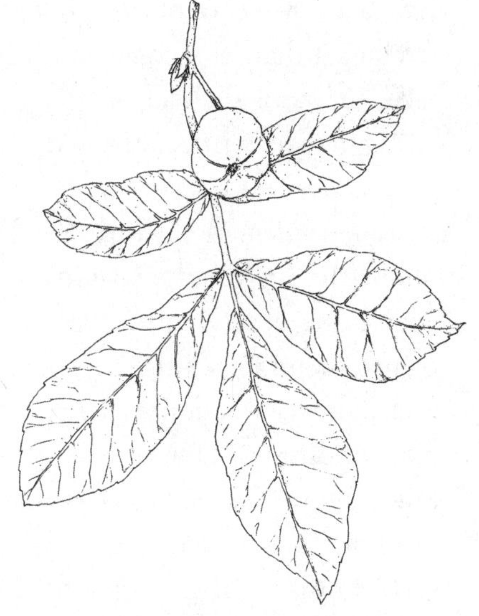

Carya ovata, **Shagbark Hickory**

are just so *easy* with nuts, and I can justify collecting bushels of them because I don't *need* to do anything with them right away.

Some nuts, like acorns, are deceptive in that way—plump and inviting, but requiring a significant investment of time, energy, and labor to leach and render them edible for humans. Nuts are a "later" food, meant to be picked at and processed leisurely over the winter months while waiting for spring to return, not a "now" food that demands my immediate attention, like mushrooms or flowers that may wilt before I can make it home.

Inevitably, that gleeful feeling vanishes the moment I sit down to begin processing them and realize what I've done—created a mountain of work a mile high, and with only myself to blame. Ask every forager you meet if they're done processing their nuts for the year, and you'll meet dozens of embarrassed people with twice as many excuses. Some of us aren't done processing nuts from *two* years ago, sometimes even longer.

That is the trick of nuts, the same one that trees have been using to ensnare overzealous foragers since time immemorial. Nuts are plentiful and free, but collecting them means entering into a contract with the tree—now you are responsible for them. You must use what you take. Take care of the gift you have been given. Our eyes are often bigger than our stomachs, or perhaps our capacity for the labor it takes to honor such a gift.

A gift. No strings attached. So what's the point from the tree's perspective?

Codependent is a word that I've heard echoing around in pop science and psychology lately—referring to a lopsided relationship where one person only gives and the other only takes, and at first glance, it might seem like trees are on the unfortunate end of a codependent relationship, especially the one they share with mammals and fungi. One of the most famous books about a tree, Shel Silverstein's *The Giving Tree*, can essentially be interpreted as an example of a codependent relationship. The boy's material desires grow in direct disproportion with the tree's needs, chipping away at each piece of her until only a stump is left. Even though she says yes each time and is "happy" to give herself away, eventually she has nothing left to give but a place to sit and rest.

The tree people have many lessons for human people, lessons about the gracious giving and receiving of gifts, and the importance of working together in community to build something good for ourselves as well as those around us.

They don't have to give us fruit and nuts to fill our bellies.

But they do.

They give us fruit and nuts, shelter, shade, even substrate. They nourish us long after death as in life. What do we have to offer trees, really?

I have several friends with whom I share a "get me back next time" system of casual reciprocity. It goes something like this: I'll pay for coffee this time, you'll get the next one. The system works when we both participate, perhaps even better than it would if we always paid for only ourselves, because it gives us another reason to spend time together in the future. It isn't an exact science; sometimes I pay two times in a row, sometimes we go to the park instead or drink coffee at each other's homes. But as long as we both continue investing in time together, the relationship will flourish.

Trees have a similar relationship with the more horizontally mobile members of our ecosphere. The gift of fruit and nuts is given freely, benefiting those who like to eat such things, but it comes with an assurance for the tree: the intentional or unintentional dispersal of seeds by the receiver. The tree will spend precious energy making flowers to attract pollinator species and creating and storing up sugar to make its product appealing enough to entice birds and mammals.

Fruit and nut production costs trees dearly, and it has almost no direct benefits for the tree itself, outside of ensuring its lineage in the case of the nuts. Trees do not eat fruit or nuts. Instead, they produce an excess of them—coordinating with other trees in the area at unpredictable intervals to flood the forest floor with nuts, more nuts than even the fattest squirrel or the most optimistic forager could ever hope to use, though many of course *will* be eaten.

The nuts are then carried off to dens by rodents, to dark places beneath the ground where the living embryos of trees will have the best possible chance to grow, if forgotten by the forager. It is estimated that squirrels and chipmunks are responsible for accidentally planting millions of trees in masting years through their collections and subsequent forgetfulness. If this is true, it would mean that the overwhelming generosity of trees and absentmindedness of squirrels have planted more trees in the history of the world than greedy oil companies hoping to offset their carbon emissions could ever begin to dream of.

Fruit is similarly carried from one place to another, sometimes stuffed into over-full cheek pockets, other times in the bellies of its grateful recipients. Subsequent deposits of seeds will typically occur elsewhere, sometimes many miles from the mother tree.

The seeds, in some cases, are more likely to be productive after passing through the digestive tract because of a process called seed coat corrosion, where the protective covering around the seed called the "coat" can be compromised and scarred by stomach acids, sometimes increasing the likelihood that it will germinate. Some trees have evolved to attract specific animal partners, self-selecting for seeds that will be optimally broken down just enough to improve germination by the digestive tract of that specific animal.

All of this to say—the giving of these gifts, fruit and nuts, benefits both parties—it feeds the receiver, and it also ensures the genetic future of the giver. It is a sobering deal to make with a plant, like caring for someone's child, but how could we say no?

IN THE CASE of the pawpaw, given the size of its fruit and its seeds, and the fact that most pawpaw groves in recent history appear to have been intentionally planted by Indigenous people, this animal partner was likely an extinct megafauna. In the absence of their original "evolution buddy," humans are now the primary animal partner for the pawpaw. We intentionally scar the seeds, soak them in potassium nitrate solutions to provide them with nutrients, and plant them in the perfect places where we know they can grow and thrive.

For some of us, the human people, we can recognize the personhood of the tree by thanking them directly, then planting some of what we harvest as a sign of respect and of gratitude. This is how one honors a gift—it is not "obligation" as the connotation of the word implies, it is an opportunity.

And, it feels appropriate to say, the tree is happy. It is a good trade, the kind where everyone feels that they've left the table better off than they were when they started. Reciprocity has been achieved.

Our customary forms of commerce, where currency is exchanged for goods and services, create an ever-growing source of supply and demand. In modern capitalism, new needs are constantly being imagined and new bodies are required to fulfill those needs. But gifts create community—the kinds of relationships that restore goodness and the ability to meet the genuine needs of those around us, and have our own needs met as a result. Gifts are the basis of a kind of social mycelium network, connecting and binding us to one another with love and responsibility all tangled up and knotted together until nobody can find where they began, no matter how hard they try, because it begins and begins again without end.

SEPTEMBER 2023

Union, Kentucky

I've just parked off the side of a pebbly driveway lined with tents, greeted by elongated, moonlit shadows of trees and the bouncing glare of flashlights at yet another foragers' gathering. By my count, that makes six . . . no, *seven* gatherings I've gone to this year alone. Seven gatherings I've taught classes at, given lectures at, led walks at. Every harvest season feels longer and longer, yet somehow, I find myself with less and less time to enjoy them.

I'm exhausted.

I make my way up to the barn at the top of the hill in front of me, which is decked out in fairy lights and smelling richly of potluck food. The stars are bright, and the air is clear and just cool enough to make me feel alert, sharpened. I can smell the deep-drinking roots of sycamores, sweet and rich, a melting smell like butterscotch. There must be running water nearby, though I can't hear it.

Out from the darkness on my left, a man's voice exclaims "you made it!" and the brief shadow of a figure with outstretched arms suddenly appears. I match the embrace with enthusiasm, immediately recognizing the owner of the voice as Paul Cencula, one half of the leadership team here at Tri Foraging. "It's great to see you!" I replied, looking left and right. "Where's Lindsey?!"

"Up at the barn," Paul replied. "There's food—most people have eaten already, but grab a plate and come hang."

"Great, thanks!"

We walk up the driveway toward the barn, and the smell of food grows stronger. I'm hungrier than I remembered. Within moments, I'm making my way through circles of chatting people, piling my plate high with a little bit of everything. Soft, gummy, spicy persimmon bread, a heaping pile of wild rice topped with some kind of curry full of wild greens that smells too wonderful to pass up, cookies and pasta salad and cornbread.

I know only three people; Paul, Lindsey, and Cole, new friends from another gathering earlier in the year, but the warmth in this place is palpable. More than anything in the world, foragers love feeding other foragers, craning their necks eagerly over your shoulder as you take a bite and then rattling off the ingredients and the process before you can even register the taste. There's a sense of pride and excitement, knowing that the person you're feeding can *really* appreciate the work and love you put into it.

I am more concerned with stuffing my face after a long drive than being social at this precise moment, but I hear Paul begin to corral everyone to one end of the barn. We find seats, some in mismatched folding and camping chairs, others on sweatshirts and blankets laid out over the gravel.

Paul clears his throat, unfolds a piece of paper, and begins to read.

"If any of your ancestors were Indigenous, they were foragers. Living off of and in harmony with this land, they are the foundation of the knowledge that we all benefit from. The Indigenous ancestors even knew that not all plants were meant for us—after all, we are not the center of the universe. These ancestors were all but pushed out of foraging spaces, as their traditional foraging grounds were stolen and walled off by colonizers. *This is sacred knowledge.*"

I am pleasantly surprised to hear Indigenous people acknowledged so immediately "from the pulpit" at a foragers' event. In my experience, Native people tend to be only occasionally acknowledged at these gatherings, and as a source of historical information or a justification for living a certain lifestyle, but rarely as individuals, only ever as concepts.

"If any of your ancestors were bonded in chattel slavery, they were foragers. Enslaved people in this country sometimes were able to make the money that allowed them to buy their way out of enslavement through foraging, and after breaking the bonds of slavery, they continued to earn a living and create a new life for themselves through foraging and selling foraged goods. In the wake of the Civil War, they were subjected to fines and punishment that made foraging all but impossible for them. *This is sacred knowledge.*"

I look around the room at faces of all shades. Some of them are nodding and humming in agreement, closing their eyes as if trying to visualize their own ancestors, to acknowledge their courage and wisdom.

"If any of your ancestors were European colonizers in this country before 1776, they almost certainly foraged for food themselves or at least ate wild food that others had foraged. Even if those European ancestors were foragers in Europe, we have different plants, mushrooms, and trees here. So who do you think taught these European ancestors how to forage here? *This is sacred knowledge.*"

Each of these ancestors has shaped the forest, left their mark on it in ways that can be recognized today. Hickory and oak orchards mark Indigenous settlements, pokeweed and sassafras line the Underground Railroad. European colonizers left marks too, many of them still visible today as persistent scars. Unchecked invasive species, deforestation, and a shameful record of theft, lies, and greed dominate this legacy.

Paul continued.

"We come to all of us, people who have benefited in so many ways from the knowledge that these ancestors learned, treasured, relied on, preserved, shared, and were punished for. This knowledge has saved lives. This knowledge is one of the primary reasons some of us are alive today. This knowledge was once one of the strongest threads that held our tribes, communities, and families together. *This is sacred knowledge.*"

The room falls silent.

"Food is not apolitical. Medicine is not apolitical. Foraging is not apolitical. Foraging is an act of defiance, and we are all food revolutionaries."

"We are choosing a better way than the one we have known."

We are here together, all forty of us, modern people from all backgrounds, woven and forged over the vast expanse of time by the dreams, love, pain, and hopes of ancient people.

I feel the words resonate and let them absorb. I forget about the ancestors too often. I still have so much to learn from them, just as I do from those around me now. There is so much to learn, too much to understand.

The ancestors still haunt us, whispering their teachings from behind the veil. Today, we are here, choosing to listen, to learn. Today, we are here, choosing to heal, together.

THE FOLLOWING DAY is busy, occupied with identification walks and foraging classes. I meet more people—Max, a master botanist and tree enthusiast who exudes effortless cool, with long hair and a crop top. James, a gentle, soft-spoken carpenter who always seems to know more than he lets on. I teach a workshop, give a talk about mushrooms, and take folks on a mushroom foray. It's dry, and the edible mushrooms aren't out in full force, but we manage to find a log of honey mushrooms, a few oyster mushrooms, and some crispy, velvet-topped turkey tail. Despite the lack of rain, people seem to be enjoying themselves, laughing, seeking, asking questions, and listening intently to my poetic waxing on about trees and their relationships with specific edible mushrooms.

"See these sycamores? They're great places to find oyster mushrooms and morels in the spring. Sometimes you can find reishi mushrooms and other *Ganoderma* sprouting out of the base of the trunk."

"These big fallen oaks and black cherry trees are perfect places to look for chicken of the woods mushrooms. If you look in the crook of a big, old black cherry, sometimes you'll find a chicken growing in the cavity."

"Normally, this pine forest is where I'd look for indigo milkcaps and porcini. It's a little late in the season now, but sometimes they'll survive the heat by taking advantage of the humidity under the duff."

The fluidity of the environment carries with it a total lack of urgency, a stretchy space where clock time seems like an afterthought. Things happen when they happen. The whiteboard with our schedule of events is more an order of operations than a prescription.

One activity catches my eye: *Nutcracking*. I think back to my unprocessed acorns and walnuts from last year, realizing that this year's nuts have already started to drop from the trees. I should have brought them with me.

***Quercus macrocarpa*, Burr Oak (nut)**

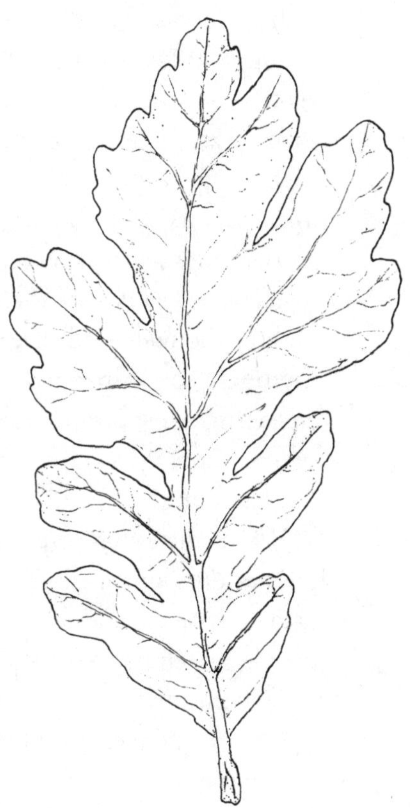

***Quercus macrocarpa*, Burr Oak (leaf)**

I wander into the barn, where an informal circle of people has manifested, all surrounding a series of small machines.

Cole pulls up a seat for me in the circle between James and Max, then hands me a small crank-style nutcracker. "You got nuts to crack?" Cole asks, returning to the pile of acorns in front of them. "Not here," I groan, disappointed that I had forgotten them. It seems that whenever a task doesn't need to be done immediately, I'm liable to keep pushing it off indefinitely.

"No worries," Cole says, "a bunch of people brought their stashes. We all have them." They gesture at the nutcracker in my hands. "That one is great for acorns, but there are others here too if you want to try them out."

It's a fantastic mix of people. Baskets and buckets and cardboard boxes overflowing with nuts are tossed everywhere—hazelnuts, acorns, hickory, walnuts. Every person in the circle is locked into their tasks, focusing on the work of their own hands, but somehow each is effortlessly in tune with the others. It is efficient, but nothing like a factory assembly line. This has a *soul*. People ease into their chosen roles as crackers and shellers, depending on what needs doing. The crackers sort their nuts into containers, then the shellers dig out the meat and discard everything else. There is a disjointed rhythm, a sort of percussion to it, the off-beat *crack* of nuts and the cascading *plunk* of nut meat dropping into plastic containers, an ensemble slipping in and out of time with one another. The song is odd, but charming.

We chatter, share tips, and crack jokes along with our growing piles of nuts and shells. The circle widens and shrinks to accommodate newcomers and relieve weary hands, and seats shift over time as people switch between roles.

"Try this one, it's better for walnuts!"

"Anyone here ever thought about making hickory oil?"

"I bought an oil expeller last year, it's out in my car! We should test it!"

"Hey, how do you guys leach your acorns?"

"Eh, depends on the acorn. White oaks go a lot faster than red, way less tannic."

"How do you tell the difference?"

"Look at the leaves. White oaks are blobby, red oaks are spiky."

"That seems like it should be more straightforward than it is."

"Did I even float these? Half of them have weevils!" Nobody knows whose nuts are whose, and nobody really cares. This space is gentle and warm and overflowing with milk and honey, the way I imagine women used to gather and waulk wool or make quilts.

Afternoon crests into the threshold of early evening, and the shadows lengthen and pirouette like a ballerina in a music box, slowing into stasis. Foraged wines, meads, and cordials are produced from campsites and cars, poured one at a time from bottle to cup, passed around like joints. We gossip, swap stories, share foraging secrets.

"This is the best elderflower wine I've ever had! How do you make it?"

"If you pick the flowers in the early morning before the dew has time to dry, the flowers have a stronger scent."

"What's the ABV on this mead?"

"Not sure. Guess we'll find out after a few glasses."

"You seriously infused *mushrooms* into vodka? How is this actually *good*?!"

"I'm telling you, chanterelles work in everything. Mushrooms just have terrible PR."

We feast—gorging ourselves on fistfuls of walnuts and leftover potluck food, on drink, on the nectar of newfound kinship. We got so much done, yet it never felt like work.

I keep looking over my shoulder, thinking something is there. Nobody ever is.

"*This is sacred knowledge,*" I hear in my head, no louder than a whisper.

The sacred knowledge of our ancestors lives on in us today in many ways. I recognize the stamp of my ancestors in the defiance of my frizzy curls, in the food I eat and the way I can't seem to pronounce the word *coffee* without adding a New York *w* in there somewhere, just like my grandma. I don't know who among my ancestors loved the natural world the way I do, but my comfort among trees and flowers and fungi feels innate, like something I could only have inherited. The part I sing with the rest of the forest is not something I could have possibly written; it must be something that I learned.

WHEN I WAS in college, one of my colleagues introduced me to a distinctive, gamified form of musical practice among Inuit women called *katajjaq*, a style some people call throat singing. It is a music, in a sense, but it is also an oral history, a way of passing the memory of past sounds from mother to daughter through generations, a sacred knowledge. Two women will face one another, practically touching, and establish a rhythmic pattern of vocalized breath. As one inhales, the other exhales, each using the cavity of the other's mouth as a resonator. The resultant sound is a disjointed, panting, staccato melody, which is prodded faster and faster as the two singers fight to keep up without running out of breath. And yet the melody becomes smoother, more unified, merging and interlocking until you can no longer identify the sonic separation between an inhale and an exhale. Though there is a competitive aspect to it, the game almost always ends with both people breaking out into peals of laughter, holding one another up as they gasp for air.

The relational, responsive function of breath in katajjaq performances never fails to make me think about humans and our relationship with trees, despite the fact that, as my friend Vivian pointed out to me, many Inuit people don't live near trees. Like the singers, trees and humans breathe only because of one another. We inhale what they exhale. When we sing, it is because a tree has done the same, it has given us a cue. As the trees breathe out, we breathe in, an eternal cycle of consuming and expelling oxygen from and for one another. The different parts of the song work only because they are sung together, resonating in tandem.

The forest is not just a place to be. It is an extension of us, a larger, more connected "us" that we step into every time we enter a forest. It is our lungs, our kidneys, our stomach, our skin. It breathes, chews, even digests for us. The very notion of the individual is almost meaningless when applied to soil and fungi, to trees and fruit, to nuts and foragers.

THE NIGHT IS lengthening, and the moon is high. Drunk on laughter (plus a bit too much of James's mead) and stuffed with at least half a pound of black walnuts, I can no longer force myself to remain awake. I wrap myself in my sleeping bag and collapse into my hammock, rocking gently as I stare up into the darkness. The stars glitter above me, blurry without my glasses but dancing white and cold in the night sky nonetheless, their ghostly waltz underscored by the rising incense of campfire smoke and sycamore.

GATHERING EXERCISE

"A GIFT FOR A TREE"

Find a tree, preferably one that looks a bit lonely.

Every day for one week, visit the tree.

Tell it about your day.

Bring a gift that you think the tree will like—a thermos of water, a handful of mulch, or tobacco if you have it.

Spend some time listening to the tree.

Breathe for the tree, breathe with the tree.

You complete each other.

Pawpaw Melomel

Makes 1 gallon

In my opinion, there are too many recipes for cooked pawpaw and not nearly enough warnings about the potential pitfalls of eating it. Every year, I watch helplessly as new foragers bring home pounds of pawpaws, cheerfully working them into breads, pies, and custards, only to hear later that their ingestion of these treats came with unexpected and devastating side effects. Without getting into details, I will simply report that I have lovingly dubbed this phenomenon "Pawpaw Toilet Trauma."

Not everyone experiences this unfortunate response to cooked pawpaw. I don't even like the *taste* of cooked pawpaw, but some people have no issue at all. It is my gentle recommendation that unless you know for sure how you react to cooked pawpaw, it is best to stick with raw fruits.

Every year (if I find enough pawpaws), I make pawpaw melomel, a sort of fruit mead. This recipe is easy to adapt to other fruits, and fortunately, it does not involve any cooking. There is a long history of pawpaw-based alcohols in North America, with records of pawpaw brandy and pawpaw beer dating back to the Civil War. Some breweries still make pawpaw booze, though usually on a small scale due to the difficulty of sourcing the fruit.

When fermenting alcohol, it is very important to consider the source of your water. Ideally, you need to use filtered, unchlorinated water to give your yeast the very best chance to survive in the fermentation environment (chlorine can stunt or altogether kill the fermentation process). If you don't have access to filtered water, simply boil tap water and let it cool to room temperature before using. This will kill bacteria and significantly reduce the amount of chlorine present in the water. This recipe calls for commercial yeast as a failsafe, but you may choose to skip the yeast and see what happens naturally. Just keep in mind that wild yeast can be unpredictable!

To obtain your pawpaw pulp, I recommend using a food processor with a dough attachment on its lowest setting (a tip I learned from Whitney over at @appalachianforager). Scoop out the flesh from the skins, seeds and all, dump it in the food processor, and spin. This will separate the pulp from the seeds, and is faster and tidier than doing it by hand. Save the seeds (keep them damp) and plant them in a new location to thank the pawpaw grove that gave them to you.

2½ pounds pawpaw pulp
1 quart raw honey, plus more for the secondary fermentation
1 sachet of Champagne or cider brewer's yeast
Funnel
2 glass carboys (or other 5-gallon container for liquid) with airlock
1 gallon filtered spring water
Cheesecloth
Bottles with carbonation-rated flip lids

For part one of your fermentation (sometimes called the "primary fermentation"), push the pulp, honey, and yeast through a funnel into a sterilized carboy. Top with the spring water. Holding the carboy top closed, swirl everything together to "stir" until the honey begins to dissolve, then add an airlock. Let it ferment for at least 2 months in a cool, dry place out of direct sunlight. The sediment should settle at the bottom.

Part two (also known as the "secondary fermentation") is called "racking," and this is where the melomel is divided into bottles to allow it to carbonate. To rack a melomel "properly," a brewer will usually use a system involving plastic tubing and another fermentation vessel. My melomels are a bit more rustic and less carefully made. I pour the liquid from the first carboy into a second carboy, which is outfitted with a funnel and a piece of cheesecloth for filtering, being careful not to disturb the "must" (sediment) on the bottom of the first carboy until about 2 inches of liquid are left. This remaining liquid and sediment are composted and not used.

When I have my second carboy full of clean, sediment-free melomel, I transfer it to my bottles, adding a tablespoon of honey to each bottle to encourage a little extra carbonation. Once bottled, I place the bottles in the same cool, dark place for a few days before transferring to my refrigerator. Always make sure you open the first one outside, or it might end up on your ceiling!

Black Walnut Hummus

Makes 1 quart

Black walnuts are nothing like English walnuts from the store. English walnuts may be mild and inoffensive, but black walnuts are *powerful*. They're hard to crack, but once you get inside, you'll be hit with a wave of bold, earthy, almost anise-like perfume. I've heard the smell and flavor of a black walnut likened to "really good paint." I think they taste the way that walking into an eccentric, absurdly wealthy person's house feels: bewildering, otherworldly, and way above my pay grade (in the nicest way possible).

The hardest thing about black walnuts is their shell, and by that I mean that black walnuts are some of the hardest nuts in North America to crack. There are about a dozen ways to get at that delicious, oily nutmeat, but few of them are easy. Some people use a hammer or a big, heavy rock to smash them, others fill bags with them and run the wheels of their cars over them, but most of my nut-head forager friends agree with me that a good-quality nutcracker will quickly become a necessity in your kitchen if you spend enough time with black walnuts. I use one called Grandpa's Goody Getter, which I reluctantly purchased for something like two hundred dollars after seeing *three* of my forager friends purchase it in a single season. It's the best nutcracker I've ever used, and it gets the nut meat out in big pieces that are easy to sort away from the shells.

Black walnuts can also be obtained in your local grocery store—just look in the baking aisle and you'll find them pre-chopped and ready to go, shells not included.

1 cup black walnuts
½ cup tahini
Juice of ½ lemon
2 garlic cloves, smashed
1 (15-ounce) can chickpeas, drained
¼ cup plus 3 tablespoons walnut oil, divided
Flaky salt and black pepper to taste

In a food processor, process the black walnuts, tahini, lemon juice, garlic, and chickpeas together on high speed until they form a paste. While the food processor is still spinning, add ¼ cup of the walnut oil a little bit at a time until the hummus is light, smooth, and creamy. Transfer to a bowl, top with the remaining 3 tablespoons walnut oil, sprinkle with salt and pepper, and serve with warm bread or crunchy veggies.

Hickory Nut Butter

Makes about 1 pint

I use the same tools to crack hickory nuts as I do to crack black walnuts, but even I have to admit, hickory nuts are a heck of a lot easier to break into. There are a lot of ways to use hickory nuts, including hickory nut milk, hickory nut ice cream, and hickory nut patties, but for those of you who don't have a lot of hyper-specific nut-crushing tools lying around, this might be a familiar recipe to ease you into the world of wild nuts (which, coincidentally, is also what I call every single one of my foraging friends).

When it comes to hickory, species matters. There are around a dozen or so different species and hybrids, all of which produce nuts that may be better for one application than another. All of them are edible, but some (like bitternut hickory, *Carya cordiformis*) are better for making oil, and others (like shagbark hickory, *Carya ovata*) have bigger, sweeter nutmeat. If you don't have access to wild hickory nuts, try substituting pecans. Hickory nuts all belong to the pecan family, *Carya*, and the flavor notes of wild hickory and pecans are similar.

After collecting your nuts and removing the husks (the green part), you'll need to float them to make sure you don't have any bad nuts. One rancid nut will ruin a whole batch of hickory nut butter. Just dump all your nuts into a big bucket of water, agitate them, and remove any floaters. You'll then take all your good nuts and crack them open, but don't

(continued)

go digging around for the meat. Instead, just boil the cracked nuts and the meat will float up to the top, saving you hours of nut-picking.

3 cups cracked, de-husked hickory nuts
3 tablespoons olive oil
2 tablespoons kosher salt
¼ cup maple syrup

Preheat the oven to 325°F.

Bring a large pot of water to a boil and dump in all your cracked hickory nuts. As the nuts boil, the nut meat will soften, separate from the shells, and float to the top. Scoop the nut meat out with a spoon and spread out on a clean, fluffy kitchen towel to dry. Discard the water and shells.

Once your nuts have dried, toss them in a medium bowl with the oil and salt and stir to coat. Dump and spread them out on a baking sheet and toast them until fragrant and just slightly darkened in color, 5 to 10 minutes. When the nuts are toasted, remove, cool, and blend them in a food processor with the maple syrup until smooth. Try it on a sandwich, in a protein shake, or as a dip for cut fruit. Store in the refrigerator in a jar for up to 1 month.

Chestnut Flour and Chestnut Pancakes

Serves 4

Chestnuts are perhaps one of my favorite nuts, and to this day are an identification I don't think I could ever forget, even if I tried. The sweet, glossy nuts are protected by a spiny casing resembling a sea urchin called a *cupule*, and be warned: after absentmindedly reaching down to pick up a few chestnuts, you'll never try *that* again. I can still feel those spikes stabbing me. Instead of fussing about with gloves and trying not to hurt yourself, you can remove the spiky bits easily by wearing heavy boots. Simply roll the nut under your foot, cupule and all, and the nut should pop out easily.

A staple in my house is chestnut flour—it's sweet, rich, gluten free, and fantastic for the hearty, carb-heavy treats we all love to indulge in during the harvest season. To prepare your chestnuts, score (cut an X in each one) and soak them in a pot of room-temperature water for a minimum of an hour. This will make them much easier to peel later on. Throw away any nuts that float during this process—these are bad nuts and will win you a quick trip down the garbage chute if Willy Wonka catches you trying to eat them.

To make the chestnut flour:

Preheat the oven to 425°F. Lay your chestnuts in a single layer on a cookie sheet, then cover lightly with foil to prevent any "popcorning." Roast for 15 minutes, then remove from the oven, flip each chestnut over, and return to the oven for an additional 15 minutes.

Let the chestnuts cool, then peel away the outer shell, revealing the yellow kernel inside. Blend the chestnuts on medium-low speed in a blender or food processor (they will be a little chunky, nothing like actual flour yet, but for now we need to increase the surface area to help them dry faster). Spread out in a thin, even layer in the dehydrator set at 120°F and leave for at least 8 to 10 hours. Blend again, this time

(continued)

on high speed, until you have a soft flour. Store in a container in the freezer.

To make the pancakes:

4 eggs

⅓ cup buttermilk

2 teaspoons baking soda

1 cup chestnut flour

Hunky pinch of salt

1 tablespoon unsalted butter, for greasing

Blueberry Ginger Syrup (page 173), maple syrup, and/or Hickory Nut Butter (page 243), for serving

Beat the eggs and buttermilk together in a medium bowl. Sift the baking soda, chestnut flour, and salt into a large bowl. Pour the wet ingredients into the dry ingredients, whisking until you achieve a smooth batter. Pour about ¼ cup of the batter at a time onto a hot, greased griddle, flipping the pancakes when the edges look delicate and lacy and bubbles appear on the surface. Serve with blueberry ginger syrup, maple syrup, and/or hickory nut butter.

Conversations

I was waiting by a tree for the leaves to fall
You came over holding flowers wanting more to love
I was waiting by that tree for the leaves to fall
Withered petals like the night in your open palms

Richard Korn,
"It's Cold Outside"

NOVEMBER 2023

Waynesboro, Virginia

Today is November first, and I have decided that I want to learn how to become invisible.

The persimmon tree stands at least twenty-five feet tall and drips with orange fruit no bigger than a golf ball, branches otherwise mostly bare. Occasionally, the orange fruits fall, one or two at a time with a gentle *whump*, louder than an acorn but softer than a walnut. It is the percussion of late autumn.

I am sitting at the base of the persimmon tree, considering how to become invisible. I am warm and solid, too perceptible. I can tell that the squirrels and birds are avoiding me, scrabbling and foraging within and around the other trees in the area. This is the only persimmon tree I can see, though I imagine she must have a mate somewhere nearby.

The Linnean binomial for the American persimmon, *Diospyros virginiana*, means "God's pear," but the tree also happens to be dioecious, a similar-sounding but unrelated botanical term meaning that sexual reproduction requires male and female flowers to be housed on different trees. Some typically dioecious plants play with the gender binary, however, like the ginkgo, a ghost of evolution that can effectively become intersex, with some branches exhibiting male characteristics and others female on the same tree.

A particularly plump, wrinkled fruit drops near my right knee. I pick it up and squeeze it into my mouth. Rich with pectin, notes of cinnamon and plum. Seedless. Mateless?

Life finds a way, I suppose.

I forgot myself for a moment. Invisible. I need to be invisible.

A scientist needs to observe the environment free of bias. Perhaps I am not a scientist in the traditional sense, but I have been observing this tree for some time, anticipating the moment where the mouth-drying astringency would be replaced with sugar. I have seen the creeping invasion of European grasses below its canopy grow tall, then wither and bend into a soft brown blanket for the fruits to land on. For months, I have waited and watched as this tree labored to ripen its fruit.

I am breathing, eyes closed, listening to the sounds around me. I imagine my body dissolving. There are very few leaves remaining on the tree, but those that cling wearily behind still manage to tap together. They sound like small birds hopping from one branch to another, if you've ever managed to pick up on the sound for yourself.

I lean back against the trunk of the persimmon tree. My hair tangles with the rough, burnt-charcoal bark, a threadlike communion as we weave together, the tree and I. Crows have complex philosophical arguments in the distance, and the Canada geese bid their farewells overhead in strict formation.

A few paces away, a sudden gust of wind activates the red-brown leaves of a chestnut oak, creating a papery cacophony as they stubbornly grip their branches. They will be one of the last to give in to winter's chilly embrace. I can hear the squirrels chasing one another down its stout trunk, competing for the remaining acorns of the season. A few fallen maple leaves blow my way, cartwheeling over my lap and tripping on my crossed legs.

"Are you okay?" I'm startled out of my deep deciduous meditation by a young man walking an elderly Labrador, which causes me to lurch forward, snapping the snagged strands of my hair and breaking my connection with the tree. "I'm fine, just enjoying the day," I chirp through a forced smile, though I do not welcome the intrusion. I don't want to explain myself or invite further conversation, and apparently, I haven't achieved invisibility yet.

An uncomfortable look flashes over his face, which I read as *Oh, this is a crazy person*, but instead he nods and says, "Alright, have a nice day."

I respond robotically with "you too" and watch him jog away, tugging the dog along as he exits stage right of my personal soundscape.

The interruption has re-alerted me to the sounds I have been attempting to listen past—cars on the highway, sirens in the distance, five men playing a pickup soccer game on the field down the street.

A scientist needs to observe the environment free of bias. But perhaps we can have a little bias today, just as a treat? I came to listen to trees and their creatures, not soccer games and freeways.

I sigh and start over. I kick my shoes off and dig my toes into the grass as far as they will go, until they touch the earth. I imagine that strings of hyphae reach out and touch the bottoms of my feet. Relax, up and up until I reach the top of my head, which leans back into the tree, completing the circuit of our connection.

I know that I'm fully imagining the sound of the hum and thrum of the tree's breath vibrating behind my back, the precise reverse of mine, breathing in carbon and expelling oxygen, but I find myself wanting to match its cadence with my own body anyway. In and out, I am circular breathing in time with the persimmon tree. I feel that it is sharing its umwelt with me as a sort of gesture of goodwill.

If a tree is breathing in the woods and you're the only one who hears it, does it make a sound?

The cool, dry air around me sharpens my nose, allowing me to pick up on the busy work of decomposition, the rapid deconstruction of plant matter that did not survive the recent frost. The molecules attack my amygdala and hippocampus, triggering a flood of autumn memories as I breathe with the tree. I fall in and out of love, revisit my childhood home, taste my mother's homemade bread, and mourn—so many half-healed losses pour out in front of me.

Today is November first, the day of the dead, Dia de los Muertos. Like the transitory celebration of life and death, the fungi responsible for creating the scents that triggered this assault of memories are an arbiter between the living and the dead. Their alchemical work of transforming dead matter into living soil makes life after death possible. They make these memories possible—a sort of echo life, calling me back until the sting of loss has diminished enough for the fondness of having-been to return.

I do not know how long I have been here, but the persimmon tree supports me in my remembering. The wind changes direction, sending its scented gifts to someone else for a time. Above me, I can hear mammalian chittering. *Perhaps I am becoming just a little bit more invisible,* I think, hopefully, but I don't want to open my eyes to disappointment. Soft *thwumps* rain around me as the creature above upsets the status quo, throwing its weight on the branches and sending persimmons flying like cherry bomb confetti.

A persimmon falls in my lap. With closed eyes, I imagine it appearing to rest in the air as it lies motionless on my invisible body. If I can make myself transparent enough, maybe it will fall through me entirely.

If a tree falls in the forest and passes right through me, do I still exist?

I open my eyes. I'm not invisible, to myself, anyway, but I appear to be invisible to the chipmunk doing parkour in the lower branches of my new persimmon friend. It zips down the trunk just behind me, then begins combing through the grass looking for spent fruit. Cheeks stuffed to the brim, satisfied that they can fit not one more fruit, it scurries off to hide its foraged treasure.

I move my hand for the first time in ages, picking up and examining the fruit in my lap. Another gift from the persimmon tree. I taste it. The familiar sweetness crosses my tongue. I spit. One large brown seed, a crescent with softened edges.

This tree continues to be full of surprises.

Seed clenched in my fist, I murmur a *thank-you* to the persimmon tree. I walk toward the South River, past the congregation of maples and oaks whispering to one another, where I know I will find a bed of soil.

It smells deeper and richer the closer I get.

I dig a hole with a stick and plant the seed, baptizing it in the soil, reminding myself that this seed, the child of the persimmon tree, holds all the genetic memories of its parent, just as all children are imprinted with ancestral memory, a gift of evolution. It is close enough to its parent that perhaps someday, their root systems will mingle with the same mycelial threads and have conversations via these buried fungal telephone lines.

In fifty or sixty years when I am no longer breathing with trees, the memory of this singular seed may hold my own ghost, invisible and watching. The years will build in jackets upon jackets of new memory, a life begun when its parent chose to entrust its future to this peculiar, ephemeral human.

Perhaps they will remember me as a silly creature, a means to an end, and likely only in autumn when the maple leaves are starting to decay, on the olfactory ofrenda of maple and oak and rotting fruit.

I can only hope that they remember me fondly, if at all.

NOVEMBER 2021
Somewhere in Michigan

Fall is a good time to be a forager. These days, my basket is always heavy. All around me are never-ending signs of plenty.

I am walking through one of my favorite forests today, a former oak savannah being slowly encroached upon by black cherry, pawpaw, hickory, some scattered stands of red and white pine, and lots of maple. The oaks are mighty, not ancient by any stretch, but much older than the other species stretching to find the light through gaps in the canopy. Showers of leaves tumble down and cover the forest floor, turning fallen trunks into indistinct lumps on the ground.

It's the perfect place to find mushrooms.

Learning to identify a *location* is just as important as learning to recognize a mushroom. Good mushroom hunters can identify mushrooms when they find them. *Great* mushroom hunters identify habitat, which enables them to find more mushrooms. Trees and topography, soil composition, elevation, moisture, temperature, time of year—all these things will get you far closer to finding the mushrooms you seek over searching blindly for shapes and colors. The process doesn't have to be as scientific as you might think. The landscape is an endless source of information, and people have learned to read it with a good deal of precision for thousands of generations.

How do you know when the rain is coming? Watch the leaves on the trees—they flip upside down before a downpour, reacting to the increased humidity in the air coupled with the breeze. How do you find the largest, juiciest berries? Look toward the sun in the lush soil, then find a place where the shade only hides the plants for a short part of the day, protecting the fruit from burning but still offering plenty of direct solar attention. How do you find due north? Look to the stars. Where are the best mushrooms? It all depends on what you're aiming to find. Each one tells a different story, each one has its own quirks.

This time of year, many people have stopped hunting for mushrooms, assuming that the season is already over. This couldn't be further from the truth. Some of the most superb edible fungi of the year only emerge once they can be assured of the promise of impending frost. Like a Minnesota college boy wearing shorts and flip-flops in the snow, they thrive in extremes (or at least claim to).

Today, I am hoping to find one of my favorite mushrooms, the polypore that proves Mother Nature sometimes saves the best for last, a fungus of many names; the hen of the woods, maitake, sheepshead, *Grifola frondosa*.

To find your own *Grifola frondosa*, you will first need to locate an oak tree. Preferably a mature one, one large enough that if you wrap your arms around it your fingers cannot touch.

When you find your oak tree, hunch over low and waddle all around the base. Look for a big bouquet of blooming mushrooms in the shape of a broody chicken fluffed out over her eggs. They may be tan, brown, gray, or a combination thereof, but you should expect to see many small, wavy fronds each branching off of a husky central stipe, exhibiting a soft white pore surface. Bisected, you might mistake it for a head of cauliflower. If you don't find a hen the first time, find another oak tree and repeat the same awkward dance, bent low, circling the base, your rear end stuck out. (A good mushroom hunter never fears a little comedy!) Do this about forty to fifty times and you're likely to find at least one. Don't get discouraged. When you eventually find a hen, you will feast for days. It's worth it, I promise. I have found them growing in urban parks, in public forests, even snuggled up beside a lone oak with no other trees around for five hundred yards.

The journey from spore to mushroom is a fascinating one. *Grifola frondosa* starts out as a tuberous structure underground called a *sclerotium*, a nutrient storage system similar in size, function, and shape to a potato. There, the fungus waits until conditions are most favorable, borrowing nutrients from its host tree, sometimes for years, at which point the mycelium will begin to put its energy toward reproduction—sending up fruiting bodies that will fling spores out into the nearby environment. The name *maitake* in Japanese translates to "dancing mushroom," purportedly because it is so exciting to find that those who happen upon it are compelled to dance with happiness. If you lie on your stomach and observe the pore surface for a while, you may be able to see a bit of magic happen when the wind changes and sends dusty sheets of white spores floating away en masse.

Grifola frondosa grows almost exclusively with oak, and it can be expected to recur on the same tree year after year. Its parasitism is oddly slow, and I have never known it to grow with trees that were not already demonstrating some kind of problem with injury or disease. Their presence is almost symbiotic, giving a tree added purpose toward the end of its natural life until it finally succumbs to the toll of living.

Maitake are fungal tree doulas, felling the mighty oak gently over the course of many years. In Japan and China, the fungus is known to reach positively enormous sizes, single fruiting bodies reaching up to 100 pounds, but in my experience here in North America, it tends to grow somewhat more modestly, with the largest I've ever found topping out at around 20 pounds or so. The hen of the woods proves itself, like so many other fungi, to be surprisingly amiable, willing to adapt and survive in a changing world, one that increasingly prioritizes human preferences.

FUNGI ARE MORE than just our neighbors, they are our cousins, in alignment with us on the tree of life as recently as 1.538 billion years ago, when they first branched off from *Animalia* to form their own taxonomic kingdom. They are estimated to be the largest kingdom of life, with the number of distinct species estimated anywhere from 2 to 12 million members, though only about 150,000 have been described by Western scientists.

Perhaps fungi know us better than we think—maybe they even *like* us. Some of them certainly seem to like the stamp we leave on ecosystems, many species preferring the disturbance created by our walking paths, our mulch, our fire, and our felling of trees. They ally with us in the production of beer, wine, bread, and cheese, and many of the unlovely ones even find homes in *our* homes in the form of mold. Some molds like the infamous black mold *Stachybotrys chartarum* may sicken you, but others, like those responsible for penicillin or cyclosporine, may save your life.

It's also possible that they *don't* like us much, which I would find somewhat understandable. Humankind's memory is short, and the memory of fungi is long. Due mostly to our destructive human tendencies, our forests have changed dramatically in recent years, to the point that there is almost no truly "native" habitat in the eastern United States. Almost all the land has been stripped, from the old-growth trees and animals down to the very topsoil itself. In the 1700s, the ubiquitous red maple *Acer rubrum* represented only 10 percent of trees in the mid-Atlantic. Now they account for approximately half. No longer are beavers abundant enough in most places to serve as the primary ecosystem engineers they once were—experts at redirecting the flow of water and creating flood conditions that restored nourishment to thirsty soils, nor has the lifegiving offering of gentle fire returned to the landscape in any significant way since the ignominious expulsion of Native people. The whitetail deer population has ballooned in the absence of predators, industry continues to slice woodlands into smaller and smaller pieces, and with it, mammalian diseases, roadkill, and forests overwhelmed with nonnative flora become the norm.

The forest I am gathering mushrooms in is fundamentally, at its core, no longer the same forest that other people gathered mushrooms in four hundred years ago. It bears the scars of colonization in every corner.

Sure, I can *imagine* this forest, make educated guesses at what it might have looked like four hundred years ago, but I would never recognize it if you showed me a photo.

Maybe the mushrooms *do* remember. The very idea of their looming judgment of our poor choices is embarrassing, to say the least. Some of them could be the offspring of the fungi that once occupied that forest, carrying memory like water through their three-dimensional hyphal tangles, or perhaps they are indeed the same fungi, sleeping underground cocooned in their sclerotium, waiting for the old forest to return, holding fast until it is safe to re-emerge.

More likely, though, when the oldest trees were felled, the fungi that had grown with them could not go on alone. Such ancient marriages tend to end only in death when one member departs. There is a strong possibility that this new forest has made some new friends, both fungal and otherwise.

I'll never know what mushrooms used to live here on this scrap of land, and it's possible that they, like so many beautiful things destroyed in this ragged country, have been purged from our collective memory. I will never know for sure if even a trace of them exists; if the mushroom I am looking at is ancestor, descendant, or new neighbor. The best I can do is offer my respects, regardless.

I get down low, stomach flat on the ground, chin pressed tight against my interlocked fingers, and I observe the magnificent fungal specimen in front of me. It's too special to pick right away. It is exactly where and when I predicted it would be, just like last year, the year before that, and the year before *that.* Easily a foot and a half in diameter, reeking of the sylvan umami that we mushroom hunters find possesses a nearly aphrodisiac quality, an intoxicant that drags us back to the same spots over and over again.

I never get tired of this feeling, the simple communion shared between the finder and the found. It is triumphant, yet sobering. The dogged persistence of the successful hunt, concluding in a moment of pure joy and wonder.

Whenever we breathe, whether it be indoors or out, we are inhaling and exhaling spores, the perpetual, tireless work of kingdom Fungi.

There is a new kind of knowing that comes with observing the behavior of an organism so different from yourself.

With my knife I saw the mushroom off at the base. I use a mushroom knife with a curved blade like a short scimitar, useful for carving out areas that may have been chewed on or in need of trimming. Its stipe is thick, fresh, squeaking as my blade severs it from the tree to which it clings. I carefully pull each frond away and examine it before placing it in my basket. Insects and debris love to burrow in the hollows and folds of mushrooms, and hens offer nearly endless opportunities for hiding places.

With the hen of the woods now a pile of "feathers" lining the bottom of my basket, I continue on, crunching through the spent leaves. *Grifola frondosa* aren't the only mushrooms peeking up from the duff in the chill today, and I'm going to have to keep moving if I'm going to find them.

The weight of a basket filled with mushrooms is addictive, a drug the likes of which nothing else can match (and I've tried my fair share). When you find your first mushroom of the day, your senses go into overdrive, all engaging at once like grand finale fireworks at a minor league baseball game. It's hard to know what to pay attention to first—the intricacies of the hymenium, the color and texture of the pileus, subtleties of the aroma or flavor, the weight and density and the lovely sounds that mushrooms make when sliced, tapped, or wiggled? Fungi are a buffet of sensory detail, an observation further amplified by the fact that mushrooms are, first and foremost, sexual organs.

The faint smell of freezer burn and florals catches me, snapping me out of my poetic fungal melodrama. *Wait . . .*

Aha! There are blewits to be had!

After falling to my knees and engaging in a bit of excited leaf-disturbing, I have found myself occupying the center of a purple fairy ring of blewits. A portal to another world, as the folklore says? Maybe so, if by "fairy world" we mean a kingdom of life that is steeped in myth, overlooked most of the time, operating almost completely outside of human rules and constructs, filled with unmatched delights as well as mortal dangers.

The wood blewit, *Collybia nuda*, or, if you like, the mushroom formerly known as both *Lepista nuda* and, perhaps worse, *Clitocybe nuda* (go ahead and laugh at the name—as noted previously, mycologists are a juvenile sort of crowd), is a personal favorite of mine, a gorgeous fungus with inrolled and sometimes wavy margins, a stocky stipe, and coloration that fades from a vivid lilac to a light tan as it ages. The gills almost always have a richer purple color than the rest of the mushroom, something that never fails to delight me when I turn one over.

Blewits mellow out beautifully when cooked, despite their questionable "past the best-by date" smell when fresh. I love to chop them up and stuff them in homemade ravioli with mozzarella, ricotta, and Parmesan, then top with a spicy red sauce, especially with fermented Calabrian chiles. Overly eager beginners may be fooled by similar-looking purple mushrooms in the genus *Cortinarius*, such as the toxic *Cortinarius violaceus*, which look passably similar in color, shape, and size, but can be easily distinguished by the presence of a cortina on the latter (a substance resembling fluffy strands of spider silk that can be observed stretching across the gills of *Cortinarius* mushrooms), as well as a significant difference in their spore prints. Blewits print whitish-lilac, while the spores of similar-looking *Cortinarius* will appear rusty-brown. Sometimes these spores get caught in the thready remnants of the cortina, which will confirm your identification on the spot.

Fungi are odd in a multitude of ways, but one of their most persistent peculiarities is certainly their propensity to defy any attempt humans have ever made to generalize them. Half-baked lies about fungi abound: "mushrooms that can be peeled are fine" (not necessarily), "don't touch any mushroom that might be toxic or you could get sick" (never happened, not even once), "every mushroom can be eaten safely when prepared correctly" (no amount of boiling, cooking, baking, or drying will destroy amatoxin, and that's just *one* example), and one of my favorites, "avoid colorful mushrooms, as they are more likely to be toxic."

Fortunately for those of us who feast with our eyes first, it doesn't take much experience to figure out just how colorful the world of edible fungi is. From the otherworldly *Lactarius indigo*, whose blue latex is immediately reminiscent of the bantha milk Luke Skywalker drinks on Tatooine to the scarlet reticulation of Frost's bolete *Exsudoporus frostii* (which may possibly be *Butyriboletus frostii* again since they can never make up their minds on where it belongs), one of the easiest tasks in the world might be to create a rainbow solely out of edible fungi. And if you did so, the blewit would be a good choice to cast in the role of "violet."

Color-linked toxicity in nature is referred to as *aposematism*, from the ancient Greek words *apo* (away) and *sema* (sign), literally "a sign to stay away." It is by no means a universal signaling system, and is primarily observed in the animal kingdom among insects, but can also be exhibited in some vertebrate organisms like poison dart frogs, which use their neon skin color as a warning to hungry animals, as if to say "keep moving, I ain't the one." There is some speculation that certain plants may also employ aposematism as a tactic to avoid being eaten, but this is limited to only a few spiny plant genera, such as agave and aloe.

But fungi? Nope. Color is an identifying feature that can be enormously valuable to consider, but it is by no means a warning system in and of itself in the fungal kingdom. Sometimes color isn't even that helpful when identifying certain species (for example, many members of the *Russulaceae* family can vary widely in cap color even within the same species). To the untrained eye, the wacky bright orange and yellow growths of the chicken of the woods *Laetiporus sulphureus* probably look like danger incarnate, while the delicate, snow-white, familiar-looking "Destroying Angel" *Amanita bisporigera* or *Amanita suballiaceae* can turn your organs into something resembling a human smoothie within a couple agonizing weeks of ingestion.

Not everything is as it appears. In Kingdom Fungi, danger is not defined by human standards of what "looks" suspicious.

Nothing in the fungal world can be assumed, no human rules are observed within. We must learn their patterns, their landmarks, inhabit their oddities and integrate ourselves into their societies if we are to understand them.

As I've said before, you have to "think like a mushroom" if you ever want to be any good at finding them. What meaning could our roads have to a subterranean organism? What hidden relationships rule the world beneath our feet?

Fairy rings, like the one I am currently occupying, are generally caused when mycelium grows in a predictable, optimized pattern within the soil. Typically starting from a point of nutrient-dense saprotrophic decay, like a buried stump or log, the mycelium of some fungi grows outward in all directions, spreading out in a pattern that appears circular from above. When the mycelium is triggered to produce fruit, typically by favorable environmental conditions, it does so in a way that maximizes the potential spread of spores, sending up mushrooms at the outer edges of the mycelial growth. This gives the appearance of a ring, which can be either small or quite large depending on how far the mycelial mat has spread.

The phenomenon is easily explainable by our limited understanding of mycelial behavior, but that does not make it any less magical. It carries with it an uncomfortably familiar sensation that one is in the presence of sentience, an alien intelligence with which we are not presently able to communicate effectively.

It's no wonder mushrooms feature so prominently in the human imagination.

I gather every other mushroom, preserving the shape of the ring but adding another layer of mushrooms to my basket. There are hundreds more in the area, growing in different patterns, some huddled and bent together like gossiping high schoolers at a lunch table, others alone, but I know that I only have immediate use for a few. My freezer is overflowing and my dehydrator is humming away—it has been a good year for mushrooms. Now is not the time to be greedy—the walk home is long and my arms are already getting tired.

The lavish blooms of fungi popping up along the walking path are a thing of luxury I did not so much as dream of seeing. Some are edible, some not so much.

Light seems softer now as the afternoon begins to wane. The fecund productivity of the fungi seems in paradox with the dying vegetation here, and although the path stretches on farther to the north, I step off to the left, heading west.

I have been known to say that a good hiker lets their eyes walk first. Those few extra seconds of visual planning can be the difference between success and disaster, or at the very least, a shameful wipeout. Two-ton boulders scattered about the steep, craggy terrain make for an extra challenging journey, especially while lugging a basket along, but with enough determination and focus, a narrow passage over and through them leading down to a gloaming depth can be identified, and it is there, at the bottom of my vision, that I must go.

Over, under, and through. The basket goes first, always, protecting the precious cargo within. To slip and fall is one thing, but spilling the mushrooms on my way down would be disastrous. Narrow tree trunks and sturdy rocks serve as railings and handholds as the basket and I make our descent.

Fwump!

I lose my footing and slide about ten feet, painfully riding along the rocks and branches on my right thigh. *That's gonna leave a bruise.* I wince, then whirl around as I remember the mushrooms. Fortunately, the basket is still safe, tucked behind a sturdy rock. *Close one,* I think to myself, relieved despite the twinge running up and down my leg.

I am very near the bottom now, perched just above it like a spider clinging to the side of a bathtub. This formation is called a "kettle," and depending on how you like to look at it, it might seem like an ancient scar, or perhaps more optimistically, a massive work of topographical art. As the shifting glaciers that dominated much of the Midwest during the Pleistocene epoch began to melt, chunks of ice broke off and landed on the earth beneath them, jumbling through the outwash and moraine in a process called *calving*. As these glacial meteors melted, they left massive craters behind, many of which filled with water and became lakes known as "kettle lakes." Most of the lakes in Michigan were formed via this process. In places where glaciers still rule the landscape, such as Iceland, the formation of kettle lakes can be observed with increasing rapidity due to the continuing effects of climate change.

***Entoloma abortivum,* Aborted Entoloma, Shrimp of the Woods**

This is a slightly more unusual kettle because it is *not* a lake. It is hollow, like a bucket that cannot hold water, eerily silent except for the occasional creaking of wizened trees. Stranger still is the warmth I feel—like hot breath, as if the kettle is the camouflaged maw of some ancient creature. What sounds like a rock falls from somewhere above into the kettle, echoing as it lands.

This place feels old, but like all places, it is still being transformed.

Transformation is not exclusive to topography, of course, and the mushroom I have braved the elements to find here is no exception. *Entoloma abortivum*, the "shrimp of the woods," is a curious autumnal fungus, the result of a parasitic relationship between *Entoloma*, known colloquially as "pinkgills," and *Armillaria*, the honey mushroom. These so-called shrimp look like whitish lumps of calcified dog poop or, perhaps wadded-up used toilet paper at first glance (appetizing, I know), but upon closer inspection, familiar fungal shapes like caps and stubby little stipes can be recognized. When cut or broken open, they may appear pink in the middle, very similar in color to a cooked shrimp. Quite like the lobster mushroom, *Hypomyces lactifluorum* (another parasitized fungus), they have a somewhat fishy flavor and smell, mild and pleasant unless you don't care for that sort of thing.

As the name suggests, for many decades it was believed that this bizarre fungus was a malformation of a single fungus, as Michael Kuo summarizes, "like a mushroom that never happened." Later on, it was proposed that *Armillaria* mycelium may be attacking *Entoloma* specimens, and later still, the inverse was postulated. We still do not know for sure who is the colonizer and who is the conquered in their relationship.

Does the framework even fit?

Was the glacier that formed this kettle the aggressor, or was it the victim of the earth below it tilting toward the sun?

Perhaps this is the wrong question to be asking. Perhaps, on the rare occasion, the whole is indeed greater than the sum of its parts.

The earth shapes the glacier, and the glacier shapes the earth, and their union brings forth a new landscape in the form of the kettle. This kettle, a legacy of ancient reciprocity, is a holding place for such entanglements, hyphal bodies grasping, clinging, and curling around one another in the dark, transforming, coalescing.

I am in the land, and the land is in me.

I scrabble back up to the basket, slide back down, and begin scooping up the blobby, white mushrooms. My fingernails and palms are stained with rich, wet soil. A slender edge of the sunset is barely visible, its crimson and orange glow diffused through the silhouettes of hairless trees dancing against it.

I lie on my back for what feels like hours with my basket resting upon my chest, watching the sky framed by the top of the kettle as twilight gives way to a sea of stars. When the incomprehensible weight of endless galaxies becomes too heavy to bear, I stand and begin to carefully ascend back up to the path, guided by the cold, ephemeral light of a porcelain moon.

GATHERING EXERCISE

"HOW TO BECOME INVISIBLE"

This is how you become invisible
Kick off your shoes. It may be cold. Embrace it.
You are becoming cold-blooded.
Set your breath like a clock in time with the pulse of the world around you.
Become soft, watch your feet blend into your surroundings.
Breathe in wind, breathe out wind.

Persimmon Bread

Makes 2 mini loaves

Ripe American persimmons are very different from the persimmons you may have tasted from the grocery store. For one, you can't really *buy* American persimmons. They have an unusual texture that can occasionally border on tacky, loaded with pectin, and when they're ripe, they're rarely shelf-stable for more than a day. Unlike most other fruits, American persimmons won't taste good when picked straight off the tree—you must be patient and wait for them to fall on their own, or mark my words: your mouth will be full of the lasting taste of chalk, bitterness, and deep regret.

Separating persimmon pulp from persimmon seeds is tricky work. The seeds are thin and oddly shaped. You can do it manually, but it's a pain. You have to press the persimmons through a sieve or a food mill, usually more than once for good measure. Sometimes you get lucky (like I have) and find a tree that produces seedless fruit. Usually, a seedless tree will be a cultivar, and often the quality of the fruit will be higher than normal. If you find that tree, make sure you take notes, share with a friend, and come back next year for more!

4 cups all-purpose flour
2 teaspoons baking soda
½ teaspoon fresh nutmeg, grated (or 1 teaspoon ground)
1 teaspoon ground cinnamon
1 teaspoon ground cardamom
1 tablespoon salt
1 cup (2 sticks) unsalted butter, softened, plus more for greasing
2 cups brown sugar
4 eggs
1 pound persimmon pulp
½ cup chopped black walnuts

Preheat the oven to 350°F and grease 2 mini loaf pans with butter.

Sift the flour, baking soda, nutmeg, cinnamon, cardamom, and salt into a medium bowl. In a separate bowl, using a stand or handheld mixer, cream the butter and brown sugar, then add the eggs, one at a time, whisking in between. Add the persimmon pulp to the mixer bowl, then add the dry ingredients to the mixer bowl, 1 cupful at a time, and mix until you have a thick, almost tacky batter. Fold in the black walnuts.

Separate the dough between the two loaf pans and bake for 40 to 50 minutes, until a butter knife poked into the center comes out clean. Serve warm. I like mine with a thick layer of cream cheese.

Pulled Hen of the Woods with Juneberry Barbecue Sauce

Serves 4 to 6

Now is the time to pull out your (possibly) last jar of lacto-fermented barbecue sauce (page 150) from earlier this year—you're gonna need it!

When you clean your hen of the woods, pull it apart one frond at a time. You'll be able to see any remaining dirt, debris, or stowaways. Ideally, you'll do this in the field where insect evictions are easier and kinder (plus it makes your basket *so* much cleaner), but sometimes this will happen in your kitchen because life happens whether we want it to or not.

For your final cleaning, rinse each piece individually, pop it in a salad spinner, and send it for a ride (or three). Rinse and spin until the water runs clear, then remove any remaining debris with a brush or carve it away. I like to make this recipe in a wok—it lets me cook everything evenly at a high temperature while preventing excessive splatter and spillage.

2 pounds hen of the woods, cleaned
2 tablespoons kosher salt
¼ cup vegetable oil
1 to 2 cups Juneberry Barbecue Sauce (page 150)
Hamburger buns, toasted (optional)
Seasonal greens (try late dandelion, lamb's-quarter, or chicory greens)

Tear the hen of the woods into small strips with your hands or two forks until it resembles pulled pork. Add the salt to a wok and turn it on to high heat. Add the hen of the woods, tossing it in the salt. Press with a wooden spoon to encourage the mushrooms to release more moisture.

When the volume of the mushrooms has reduced by about half and the mushrooms begin to stick to the bottom of the wok, add a generous glug of vegetable oil. Fry the mushrooms until crispy, then add the barbecue sauce, stirring and tossing the mushrooms around to coat evenly. Stop stirring and let the sauce caramelize around the mushrooms for 1 minute, then remove from the heat. Serve the barbecue hot on toasted buns with seasonal greens.

Half-Spicy Blewit Pasta

Serves 2 to 4

I always look forward to the end of mushroom season when the blewits are out in full force. Blewits, like so many of the mushrooms I've fallen in love with, are not particularly sought after for their culinary merits. Even so, I find them delicious—stunning and purple, beautifully meaty and firm with a texture that yields gracefully to the gentle ministrations of a spicy cream sauce. There is a finality to blewits, a "last call" of the late autumn mushroom offerings as we must finally begin to bundle up and pack our baskets away for the long winter ahead.

1 pound blewit mushrooms, sliced ¼ inch thick
1 tablespoon kosher salt, plus more for cooking the pasta
2 tablespoons olive oil
2 shallots, diced
2 tablespoons fermented Calabrian chile paste (can substitute other chile paste, depending on your heat preferences)
5 garlic cloves, minced
1 (16-ounce) can whole San Marzano tomatoes
¼ cup heavy cream
½ cup grated Parmesan cheese
1 cup loosely packed fresh basil, torn
1 pound dried pasta
1 cup reserved pasta water
Flaky salt, for finishing

(continued)

Dry sauté your mushrooms by adding them in a single layer to a very hot, heavy-bottomed pot with a good dash of salt and no oil. Allow them to form a light crust before flipping over, then allow them to reduce in size by half. Add olive oil and stir to coat the mushrooms, then add the shallots, chile paste, 1 tablespoon of salt, and garlic and sauté together until the alliums are fragrant and slightly browned and the shallots are translucent, 3 to 5 minutes. Add the tomatoes and break them up into small chunks using the back of your spoon. Lower the heat to a simmer. Stir occasionally until the sauce is thick and bubbling. Add the heavy cream and ¼ cup of the cheese, plus a big handful of basil. Stir, let the mixture simmer until bubbling, and then turn off the heat.

Prepare your pasta of choice in salted boiling water, reserving 1 cup of the pasta water before straining. Add the pasta water to the sauce, stir, then add the pasta to the sauce and toss to coat. Serve with the remaining ¼ cup cheese, some flaky salt, and the rest of the basil.

Spicy Shrimp of the Woods Onigiri

Serves 3 to 6

Onigiri are Japanese sticky rice balls stuffed with an endless variety of fillings, usually pickled fruit, vegetables, seafood, or other proteins, then pressed into triangular shapes and wrapped in nori (seaweed). It's almost impossible to get bored of onigiri, because you can fill them with almost *anything*. Onigiri are like very cute sandwiches, and they couldn't possibly be easier to make. I bring a lot of onigiri to work with me, since they are portable, filling, self-contained, and don't require utensils or a microwave, and if I ever get bored of a flavor, I can always switch it up. I tend to experiment and use whatever mushrooms I happen to have available, but mushrooms that have a natural seafood-like flavor such as shrimp of the woods, lion's mane, or lobster mushrooms are the best for this. You can also try making some with your lacto-fermented black trumpets (or other mushrooms) from earlier in the year (see page 146)!

Onigiri works best when made with a short-grain rice, in particular sushi rice. When rinsing your rice, don't let the water run totally clear—leave some starches behind so that the rice stays sticky and holds together after it is refrigerated. When you are working with the cooked rice to form your balls, use wet hands to keep the rice from sticking to them. I keep a small bowl of water next to my workspace so I can dip my hands anytime I need to. If you are making your onigiri for later, you may want to wrap them tightly in beeswax wrap and pack your nori separately, otherwise the nori will get soggy (which not everybody likes).

2 cups short-grain rice
2½ cups cold water
2 tablespoons rice wine vinegar
1 tablespoon mirin
2 tablespoons furikake rice seasoning (optional)
1 tablespoon kosher salt
1 cup shrimp of the woods, cleaned and chopped into ¼-inch pieces or smaller
2 tablespoons sriracha, or to taste
2 tablespoons Kewpie mayo (can substitute with plant-based or another mayo), or to taste
2 sheets nori seaweed, each cut into 4 equal pieces

Using your rice cooker's insert or a heavy-bottomed pot, rinse the rice with cold water one or two times, or until the water does not run entirely clear but still appears cloudy. Drain and add the 2½ cups cold water. Soak for at least 20 minutes before cooking (this will help your rice to cook more evenly).

Add the vinegar, mirin, and furikake to the rice. If you are using a rice cooker, cook according to your rice cooker's instructions for white rice. If you are using your stovetop, cover and bring the rice to a boil over medium heat. Once boiling, reduce the heat to low and simmer for 12 to 15 minutes, until the rice is no longer crunchy in the middle. Keep the lid on until the rice is fully steamed.

(continued)

Meanwhile, heat up a cast-iron pan over medium-high heat. Add the salt, then the shrimp of the woods. Unstick the mushrooms with water as needed, pressing with a spatula to release moisture and turning them over until a crust forms on both sides. When the mushrooms have reduced in size by about half and have developed a golden brown crust, remove from the heat and set aside.

As the mushrooms cool, combine the sriracha and mayo in a medium bowl and whisk together. You may wish to adjust the ratio of ingredients, depending on your preferred spice level. When the mushrooms have cooled enough to handle, toss in the bowl with the condiment mixture to coat.

When your rice is fully steamed, fluff with a fork or a rice paddle and set aside to cool. When the rice has reached a comfortable temperature to handle with bare hands, move to your workspace with the mushrooms, nori strips, and a bowl of water.

To shape the rice balls, wet your hands and make a ball of rice about half the size of a tennis ball. Form it into a semi-round shape with a hollow crater on top, almost like a volcano, then fill the hole with about 1 to 2 tablespoons of your shrimp of the woods mixture. Cover with more rice, then use your fingers to form the ball into a triangular shape. Don't be afraid to use some pressure, as you will want the ball to hold its shape.

To finish, take a strip of nori and wrap it around the rice ball. You can make more mayo/sriracha sauce for dipping or eat it as is. Store in beeswax wrap or plastic wrap in the refrigerator and enjoy within 2 days.

Epilogue

O perpetual revolution of configured stars, o perpetual recurrence of determined seasons, o world of spring and autumn, birth and dying!

T. S. Elliot,

The Waste Land and Other Poems

Once again, winter has caught me in no small way by surprise. Her approach, as always, is slow and mostly silent, marked by subtraction rather than addition. I rarely seem to notice the moment when the last birds have evacuated to the warmer south, fattened on the bounty of berries and seeds of autumn, yet I feel it when they are gone. The absence of their foraging songs leaves a gasp in the taskscape.

The large, fluffy snowflakes collecting on my hair are a stark reminder that the proverbial shop is closing. I press my hands a little deeper into my fleece-lined pockets, forcing the last of the year's warmth into my fingertips.

I used to hate the winter solstice. The loneliness, the silence, the fruitless howling of the wind as it spilled over the edge of the world.

The darkest day of the year.

Selah.

This winter feels different than other winters.

The ancient Greek Stoics were highly concerned with the concept of eternal recurrence, also sometimes called eternal return, in which time repeats itself in an infinite loop and all events, both small and large, will repeat themselves for all of time immemorial. Put simply, "history repeats itself," but perhaps a bit more literally than your history teacher took it.

Let me assure you that I have no intention of inflicting an unwanted lesson in philosophy upon you (we've all suffered enough). However, through observing the seasons as they fluctuate and change, I cannot help but recognize themes of recurrence all around me. My sorrow at the departure of the birds is tempered considerably because I believe they will see fit to return to my backyard in the spring. I recognize the sleeping and waking cycles of the plants, animals, and fungi that I share my world with, and I know that they will grow, give birth, and die. I assume that the sun will rise in the morning; the only reason to believe this is that it has happened before.

There is a certain comfort in the notion that all is now as it shall be again, and yet my fear in these unusual times of war, environmental degradation, climate change, and impending ecological disaster is that all is now as it *never* shall be again.

I don't know how to fix all of it.

But maybe I don't have to.

This year, my boot prints marking my departure from home are not an escape from some shadowy misery, they are just part of the journey that will eventually lead me back, frigid hands full of winter mushrooms and pockets stuffed to bursting with acorns and rose hips. Long gone are the old boots stuffed with plastic bags—my socks are miraculously dry, snuggled into warm mukluks lined with rabbit fur. A gift, one that fills me with gratitude each time I slip them on.

I used to feel alone, even when I was surrounded by people. Now I feel like I belong to others, even when I am entirely alone. There is something peaceful about it, something warm. Accepting.

My home is warm and inviting. I see my cats staring out the window, one orange and one black lump fixated on the cardinal flirting with them in the bushes, freshly dusted with new snow.

Our ancestors used to spend their winters gathered together, surviving off what they had spent the warmer months growing, hunting, and collecting. They understood that self-sufficiency was an impossible goal, that an individual was only ever as strong as the community that claimed them.

Waiting for the sun is better when surrounded by friends.

Knowledge is useful, yes, and it can get you pretty far. Diagrams of plants, books of facts, "eat this not that," "here's how to grow taller corn," "here's how to start a fire," "here's how to build a shelter."

But knowledge won't lend you a cup of sugar, put its hand on your plow when your knee starts acting up, or trade canned goods for fresh eggs.

Wisdom guides us toward the future. It instructs us to create community. Trust. Not to hoard, but to share. To build, together.

Wisdom urges us to gather.

Many cultures tell stories in the winter. The Ojibwe refer to the entire winter season as "a time for stories." It makes sense from a practical perspective: strenuous tasks and preparations keep us busy throughout the growing seasons—humans throughout history have followed Mother Earth's lead and rested on her schedule. But in the winter, we settle. We park ourselves, even when our hands stay busy. When we're not rushing around to get things done, we have more time to listen.

But there is something else to it, a deeper magic. The mind is quieter with the absence of environmental activity, the invitation to sit by the fire somehow becomes more intoxicating. Winter's darkness makes the mind prickle with imagination, and the fire sends shadows dancing on the snow like ancient, nameless spirits.

I think about lighting a fire in the backyard for us to enjoy, but the urge to curl up in the comfort of the indoors is too strong.

I want to greet you with coffee and warm spicebush cookies, with blankets and softness and no expectations.

Baskets of acorns, hickory nuts, and walnuts are placed on the floor, propping up a precarious stack of cushions and blankets, waiting for you to arrive and claim your comfort. A couple of you will be drawn to the pot of chili cooking on the stove, digging through my spice cabinet to make adjustments. Some of us will crack nuts in a circle, losing hours in the simple meditation of work done together. One of you will take it upon yourself to fill the room with music. A few of you will sit around with your own projects, knitting or making cordage, and perhaps some of you will have idle hands but animated conversation, tending in your own way to the social garden of this space.

Winter, after all, is a time for telling stories.

Selah.

I hope that you will feel comfortable here, in this home I have built. It is the sort of place where you don't think to knock, to ask for what you need, instead going to the cabinets or the fridge and helping yourself. I have not gone to any pains to make the place appear tidier for your benefit, nor have I so much as washed a single dish from breakfast. This isn't because I don't care enough about you to sanitize my chaos—it's because I care enough to let you be with me and to be with you, bare-faced and wholly myself.

I will abandon my quest of perfection in pursuit of a greater good, unfiltered and unwashed time, the collision of my spirit with yours.

The floor will be littered with nutshells after a while, and despite the abundance of hands in the kitchen, the cornbread will still somehow burn on the edges. There will be coffee spilled down the front of shirts, cookie crumbs caught in beards and couch cushions. Laughter and gossip and dreams of the future will melt together like clouds in the air as hyphal threads of half-remembered stories wend through the room, a network connecting us all.

When the bones in our fingers have grown too tired to continue working, we will rest together. Bowls of steaming chili topped with crumbly, overbaked cornbread will pass from hand to hand, and treasures you all have brought from your own homes turn this simple meal into a feast—spicebush cookies and sticky berry jam, persimmon bread with fragrant hickory butter, polenta cake and mugolio, all manner of treasures preserved from the abundance of warmer months.

I didn't ask you to bring anything, but we will be all the more well-fed for it.

It's funny how food multiplies when it is shared.

We are all, each of us, essential to the task at hand, to this gathering.

You are worth being known.

You are worth giving gifts to.

Belonging.

We will tell stories we have all told a thousand times, and laugh at the funny bits like we're hearing them for the first time. For a while, every sentence will begin with "remember that one time when?" and end the same.

Circles of stories, woven in the tapestry of winter, strung with the threads of human voices. Pictures are forming.

Outside, the wind is howling. Snow is deepening, filling in the tracks I left before the darkness crept in.

The spirits are dancing.

Perhaps the world will end here, in the fiery ekpyrosis predicted by the Stoics, or maybe it will go out with a whisper as it slowly forgets how to perform the essential task of wintering, spinning itself into the hot loneliness and weariness of overproduction.

If the world ends at this moment, you and I will be here at my table together, savoring the last toothsome bite, believing in and awaiting the Light that threatens never to return.

When it is done and the world is ash around us, we will pass the dying flame from the last remaining coal, candle to candle, wick to wick, calling one another back to the fire where it all began. To warmth.

We will share what we have.

We will give.

We will gather.

It is cold outside, for now.

It is cold outside, and that means it is time to tell stories, the ones that will coax the Earth from her dreaming, that will make her remember us.

In the story I am telling you, a story that will never end as long as you hold its fire in your palm, I will take your hand and lead you to the fallen logs where the mushrooms grow.

I will whisper your name to them, so that they may know you as they know me.

GATHERING EXERCISE
"PASSAGE"

Use your hands to make something
Perhaps
A crocheted potholder
A drawing of a snail in the corner of a fresh, clean notebook
A loaf of bread
A bookmark
A cup of coffee
A song
Make it mindfully. Beautifully.
Put your heart into it, whether it takes you ten minutes or a week.
Then you must take this thing you have labored over
this gift of your hands
wrap it up, if you can
and give it
Away.
Not to someone you already know, a friend, a family member, a partner
but someone unknown to you.
Exchange names, and remember theirs as one remembers the taste of honey.
Repeat these steps as often as necessary until you have given yourself away to the whole world a hundredfold.

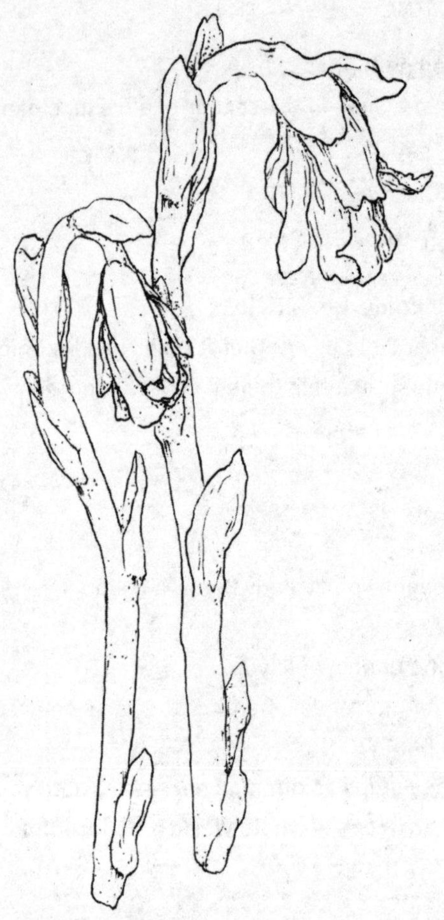

Monotropa uniflora, Ghost Pipe

Acknowledgments

I COULD NOT HAVE WRITTEN THIS BOOK without those varied and wonderful organisms that have taught me how to be a more gracious, forgiving, reciprocal member of our society and ecosystem, despite my many glaring flaws and shortcomings. The mushrooms, the plants, and the animals have been my most constant and effective teachers, and their wisdom continues to humble me. Thank you to all my nonhuman teachers.

This book would not be possible without the Indigenous knowledge bearers on this continent of Turtle Island who, against all odds, have preserved and continue to pursue highly advanced ecological science, and I am particularly grateful to the many incredible people who have chosen to share their knowledge with me.

Thank you to my parents, Mark and Julie Cerberville, and to my two younger brothers, Christian and Zachary.

Thank you to Timothy Cronican, my big brother, best friend, and advocate.

To my sisters, Kim Fenton, Kelly Muzzin, Andy Cronican, and to Harper and Xander Cronican, the OG niblings.

Thank you to Rudy Colantonio, The Analyst, for being the best hype guy anyone could ask for.

Thank you to Dave and Jaye Eisinger for always being in my corner.

Thank you to Pascale and Troy Hollings for inviting me to invade their peace and move onto their farm in my eyesore of an RV for an entire year.

Thank you to the whole motley crew at the Broad Ripple Books & Brews.

To Frank Felice, my teacher and mentor and someone who I have been privileged to call my dear friend for over a decade. You have believed in me every step of the way, especially when I felt I didn't merit that belief. There is not one speck of what I have today that I do not in some way owe to you.

Thank you to Lisa Coons, one of my most inspiring teachers. You deserve everything down to the marrow of this beautiful, wild world, and no, you're never overstepping.

Thank you to all my professors at the University of Virginia, especially Matthew Burtner, for your flexibility and encouragement as I juggled being a student with writing this book.

Thank you to my agents, Wendy Sherman and Callie Deitrick, to the impeccably organized and ever-encouraging Sarah Haugen, my editor at Harper, and to Ezra Kupor and JB Douglas for their invaluable feedback on the recipes contained in this book.

Thank you to my expert editors, Samuel Thayer (plants), Ariel Bonkoski (fungi), and Vivian Yéilk' Mork (Indigenous knowledge), three brilliant naturalists whose work I have long admired. I am confident that this book is going into the hands of hungry readers with the best possible information thanks to your careful fact-checking and impeccable attention to detail.

Thank you to my incredible illustrator, Madison Memering—a brilliant and thoughtful naturalist in her own right who perfectly honored the detail, beauty, and imagination of the incredible organisms I wrote this book for and about.

Thank you to the many amazing members of the foraging and mushroom communities I haven't already thanked who have offered me friendship, guidance, support, and countless snacks over the years. I am especially grateful to Alexis Nikole Nelson, Betsy Hinze, Sasha Ospina, Whitney Johnson, Lindsey and Paul Cencula, Daniel Liberson, Debs Gardner, Maxwelle McCormick, Megan Weatherall, Gordon Walker, and so many more. Thanks to all of you fellow ecologists and forest people who share my eyes; I do not have to live, as Aldo Leopold once said, "alone in a world of wounds."

Thank you to Katie Hamilton and Linnea Geno. No matter how we have changed and grown, whenever we are together I still see us exactly as we were so many years ago.

Finally, thank you to all of you, whether you decided to pick up this book because you wanted to hear what that loudmouth mushroom nerd from the Internet had to say, or maybe you just needed something to bring with you to the beach. I hope this book has been everything that I dreamed it could be for you. I'm grateful to you for spending your time with me.

Thank you all. Thank you for everything. I couldn't have done it without you.

Alphabetical List of Plants and Fungi

ALLIARIA PETIOLATA (Garlic Mustard)

FLOWERS AND FRUIT

- Small, white, 4-petaled flowers in clusters.
- Slender seed pods (siliques) turning brown in fall.

LEAF FEATURES

- First-year rosettes with kidney/heart-shaped leaves and softly toothed margins.
- Second-year triangular leaves and softly toothed margins.
- Leaves are slightly greasy and smell strongly of spicy garlic.

STEM FEATURES

- Erect, ridged stems 1 to 3 feet tall in second year.
- Basal rosettes first year, then flowering stalks.
- Flexible, deep taproot, staining purple at the root/stem attachment.

GROWTH HABIT

- Biennial, forms dense stands, outcompeting native flora.

HABITAT AND GROWING CONDITIONS

- Moist, rich, partial shade in woods/edges.
- Invasive in North America.

NOTABLE IDENTIFYING CHARACTERISTICS

- Garlic odor when crushed.
- White flowers, scalloped leaves.
- Rapidly spreads in forest understories.

Recipe featuring *Alliaria petiolata*: Lacto-Fermented Garlic Mustard Pesto (page 78).

ALLIUM TRICOCCUM (Ramps, Wild Leeks)

FLOWERS AND FRUIT

- White spherical cluster on a leafless stalk after leaves die back.
- Round seed capsules with glossy black seeds.

LEAF FEATURES

- Broad, elliptical leaves (1 to 3 per bulb); strong onion-garlic smell.

STEM FEATURES

- Ephemeral leaves emerge directly from bulbs below ground in spring.
- The base of each leaf may, at the point of attachment, appear either reddish, slightly purple, or white.

GROWTH HABIT

- Perennial from bulbs, forming dense stands.
- Leaves wither by summer; flowering stalk appears afterward.

HABITAT AND GROWING CONDITIONS

- Rich, moist deciduous forests.
- Often in large colonies in shady areas.

NOTABLE IDENTIFYING CHARACTERISTICS

- Strong onion-garlic aroma.
- Broad leaves emerging early spring, disappearing by summer.

Recipes featuring *Allium tricoccum*: Curried Pheasant Back Vegan White Chili (page 72), Fiddlehead Summer Rolls (page 74), Ramp Leaf Compound Butter (page 76), Ramp Leaf Salt (page 77), Morel Risotto with Saffron, à la Debs (page 80), Wildcrafted Sofrito (page 71).

ALLIUM VINEALE (Crow Garlic, Field Garlic)

FLOWERS AND FRUIT

- Umbel with aerial bulbils (bulb clones appearing at the top of the plant in summer) and/or small greenish-white to pinkish flowers (summer).

LEAF FEATURES

- Narrow, hollow, cylindrical leaves; strong garlic-onion odor.

STEM FEATURES

- Smooth, upright scape, may be curly or straight.
- Arises from an underground bulb, usually small and smelling strongly of onion.

GROWTH HABIT

- Herbaceous perennial forming clumps.
- Spreads via aerial bulbils and underground bulbs.

HABITAT AND GROWING CONDITIONS

- Fields, lawns, roadsides.
- Invasive in North America.

NOTABLE IDENTIFYING CHARACTERISTICS

- Hollow, dark green scapes.
- Clustering habit, small underground bulbs.
- Strong onion-garlic smell.

Recipes featuring *Allium vineale:* Miso Soup with Enoki (page 40), Wildcrafted Sofrito (page 71), "Cheesy" Field Garlic Popcorn Dust (page 44), Curried Pheasant Back Vegan White Chili (page 72), Fiddlehead Summer Rolls (page 74).

AMANITA MUSCARIA VAR. *GUESSOWII* (Eastern Fly Agaric, Yellow Muscaria)

CAP

- Yellow to orange, generally with white warts (veil remnants).
- Margin striated.
- Domed when young, becoming broadly convex with age.

UNDERSIDE FEATURES

- Crowded white gills, free to narrowly adnate.

STIPE FEATURES

- White to yellowish, bulbous base with volva remnants.
- Slightly shaggy.

GROWTH HABIT

- Solitary or scattered, summer/fall.
- Mycorrhizal with conifers/hardwoods.

HABITAT AND GROWING CONDITIONS

- Common under pine, birch, and oak in eastern North America.
- Prefers moist, forested areas, forest edges, and disturbed grasses.

SPORE PRINT

- White

NOTABLE IDENTIFYING CHARACTERISTICS

- Lemon yellow to burnt orange color with white warts that can be picked off.
- Bulbous base, universal veil remnants, shaggy stipe.

Edibility note: Edible only with specific preparation methods; contains toxic and psychoactive compounds (no featured recipes).

AMELANCHIER (Serviceberry, Shadbush, Juneberry) *(Genus)*

FLOWERS AND FRUIT

- Flowers: White, 5-petaled in showy racemes, early spring.
- Fruit: Small, round berries ripening to purple-blue, sweet and edible.

LEAF FEATURES

- Oval to elliptical, finely serrated margins.
- Young leaves may be bronze-tinged.

STEM FEATURES

- Smooth to slightly furrowed bark; young twigs grayish or reddish-brown.
- Can be multi-stemmed or single trunk.

GROWTH HABIT

- Small deciduous trees or large shrubs (6 to 25 feet).

HABITAT AND GROWING CONDITIONS

- Edges of woodlands, forest understories, ornamental plantings.
- Well-drained, slightly acidic to neutral soil, sun to partial shade.

NOTABLE IDENTIFYING CHARACTERISTICS

- White spring blossoms, edible purplish berries, smooth gray bark.
- Attractive ornamental and wildlife food source.

Recipes featuring *Amelanchier*: Lacto-Fermented Juneberry Barbecue Sauce (page 150), Pulled Hen of the Woods with Juneberry Barbecue Sauce (page 266).

APIOPERDON PYRIFORME (Pear-Shaped Puffball)

CAP

- Entire fruiting body is pear-shaped, 1 to 2 inches. May be brownish or whitish.

UNDERSIDE FEATURES

- No distinct underside; forms a hole at the top for spore release when mature.
- White interior when young, turning brown/powdery. Specimens are only edible when white all the way through.

STIPE FEATURES

- Narrow base (pseudo-stem), part of the same body.
- Grows in dense clusters, typically sprouting from dead wood.

GROWTH HABIT

- Clustered on decaying wood (logs, stumps).
- Late summer to late fall, extending into early winter.

HABITAT AND GROWING CONDITIONS

- Common in damp, shaded forests on rotting wood.
- Grows in dense clusters.

SPORE PRINT

- Interior spores are olive to brown at maturity.

NOTABLE IDENTIFYING CHARACTERISTICS

- Pear-shaped, clustered on wood, white interior turning brown.
- Releases spores from a small hole at the top when mature.

No featured recipes.

ARMILLARIA MELLEA (Ringed Honey Mushroom)

CAP

- Honey-colored to yellow-brown, often with small dark scales, darkening with age.
- 2 to 6 inches wide, slightly sticky when fresh.

UNDERSIDE FEATURES

- White to pale yellow gills, attached to the stem.
- Ring (partial veil) typically present on younger specimens.

STIPE FEATURES

- White, membranous ring on stem.
- Bulbous or fibrous, darkening at base, often coming to a point.

GROWTH HABIT

- Clusters on or near wood (stumps, logs, living trees).
- Parasitic on trees, using rhizomorphs (finger-like black appendages) to dig into inner tree bark.
- Late summer/fall fruiting.

HABITAT AND GROWING CONDITIONS

- Parasitic/saprotrophic on hardwoods.
- Common in temperate forests.

SPORE PRINT

- White.

NOTABLE IDENTIFYING CHARACTERISTICS

- Honey-colored cap, ring on stem.
- Grows in large clusters.
- Mycelium is bioluminescent (as with all honey mushrooms).

Recipe featuring *Armillaria mellea*: Honeyed Honeys (page 219).

ASARUM CANADENSE (Wild Ginger)

FLOWERS AND FRUIT

- Flowers: Brownish-purple, smelling of rotting meat, bell-shaped at ground level (spring).
- Fruit: Small, fleshy capsule splits open at ground.

LEAF FEATURES

- Large, heart/kidney-shaped leaves, glossy on surface.
- Leaves in pairs, low to the ground.

STEM FEATURES

- Mostly subterranean rhizomes; above-ground stem length varies based on available light and amount of leaf litter.

GROWTH HABIT

- Low-growing perennial, forming dense patches via rhizomes.
- 4 to 8 inches tall.

HABITAT AND GROWING CONDITIONS

- Rich, moist woods in shade.
- Acidic to neutral soils, leaf-litter layer.

NOTABLE IDENTIFYING CHARACTERISTICS

- Gingery aroma from rhizomes when crushed.
- Heart-shaped leaves.
- Edible in moderation with caution, contains aristolochic acid.

Recipe featuring *Asarum canadense*: Wild Blueberry Ginger Cocktail Syrup and Gimlet (page 172).

ASIMINA TRILOBA (Pawpaw)

FLOWERS AND FRUIT

- Flowers: Maroon, drooping, bell-shaped, 3-petaled, appearing in mid-spring.
- Fruit: Greenish-yellow oval (2 to 6 inches), sweet, custard-like flesh; large brown seeds. Ripe late summer/fall.

LEAF FEATURES

- Large and simple, widest near tip.
- Leaves smell pungent when crushed, like composted horse manure.

STEM FEATURES

- Bark thin, gray-brown; young twigs greenish-brown.
- Often forms clonal thickets via suckers.

GROWTH HABIT

- Small understory tree/shrub, typically no more than 30 feet tall.
- Spreads by root suckers.

HABITAT AND GROWING CONDITIONS

- Rich, moist, well-drained soils in shady forest understories.
- Bottomlands, stream banks in eastern North America.

NOTABLE IDENTIFYING CHARACTERISTICS

- Tropical-looking leaves; unique oblong fruits.
- Maroon spring flowers, strong fruit aroma/flavor.

Recipe featuring *Asimina triloba*: Pawpaw Melomel (page 240).

BOLETUS EDULIS CLADE (Porcini, Penny Bun, Ceps, King Bolete Group)

CAP

- Brownish, matte or slightly greasy, 3 to 10 inches wide, but species around the globe may vary in size and color from very pale tan or whitish to a purplish red.
- Margin often paler when young.

UNDERSIDE FEATURES

- White to pale yellow pores, turning olive with age.
- Pores small and spongy.

STIPE FEATURES

- Thick, club-shaped, faint reticulation (net pattern) on upper portion.
- White flesh does not bruise blue.

GROWTH HABIT

- Solitary to scattered, summer/fall.
- Mycorrhizal with conifers/hardwoods.

HABITAT AND GROWING CONDITIONS

- Temperate forests worldwide. Mycorrhizal partners vary depending on location.
- Prefers slightly acidic to neutral soils.

SPORE PRINT

- Brown.

NOTABLE IDENTIFYING CHARACTERISTICS

- Bulbous stem with net-like reticulation.
- Non-bruising white flesh.
- Highly prized edible, tastes sweet/mild in the field as opposed to similar-looking bitter species.

Recipe featuring *Boletus edulis* clade: Umami Bolete Dust (page 177).

CALVATIA GIGANTEA (Giant Puffball)

CAP

- Entire fruitbody is a large, white, lumpy ball, often much larger than a basketball.
- Has a thick "skin" that can be peeled away in younger specimens.

UNDERSIDE FEATURES

- No gills or pores; interior is solid white with a Styrofoam texture when young, softening and turning yellow/greenish as spores mature.

STIPE FEATURES

- Essentially stemless; a large globe attached to ground by a small root-like base (I call it the "mushroom tootsie").

GROWTH HABIT

- Single or in small groups in fields, meadows, lawns.
- Late summer to fall.

HABITAT AND GROWING CONDITIONS

- Prefers open grassy areas with rich soil.
- Often reappears in same location yearly.

SPORE PRINT

- Interior spores are olive to brown at maturity.

NOTABLE IDENTIFYING CHARACTERISTICS

- Very large, white spherical body.
- Edible when interior is pure white (immature).
- Turns powdery inside at maturity.

Recipe featuring *Calvatia gigantea*: Giant Puffball Pizza (page 212).

CANTHARELLUS SPP. (Chanterelle) *(Genus)*

CAP

- Funnel/vase-shaped, tones ranging from pale yellow to deep orange.

UNDERSIDE FEATURES

- Blunt, forked ridges (not true gills).

STIPE FEATURES

- Same or slightly lighter color than cap, typically dense. Can be peeled into strips like string cheese.

GROWTH HABIT

- Solitary or scattered in leaf litter, oak/hardwood forests. Often follows streams and tree roots.
- Summer to early fall.

HABITAT AND GROWING CONDITIONS

- Mycorrhizal with hardwoods.
- Warm, moist conditions, rich woodland soils.

SPORE PRINT

- Pale cream.

NOTABLE IDENTIFYING CHARACTERISTICS

- Fruity/apricot aroma.
- Most species bright yellow with decurrent ridges instead of gills.

Recipes featuring *Cantharellus* spp.: Chanterelle Peach Pie with Basil (page 144), Thai Chanterelle Red Curry (page 148).

CARYA OVATA (Shagbark Hickory)

FLOWERS AND FRUIT

- Flowers: Inconspicuous catkins (male) and spikes (female).
- Fruit: Fat, spherical husk splitting in four sections, revealing a hard-shelled nut.

LEAF FEATURES

- Pinnately compound, serrated leaflet with 5 sections, terminal leaflet largest.
- Bright green above, paler beneath.

STEM FEATURES

- Bark in long, shaggy strips, may appear to be "shedding."
- Twigs stout, large terminal buds with loose scales.

GROWTH HABIT

- Large deciduous tree, 60 to 80 feet, can exceed 100 feet.
- Straight trunk, oval crown.

HABITAT AND GROWING CONDITIONS

- Moist, well-drained soils on slopes/valleys.

NOTABLE IDENTIFYING CHARACTERISTICS

- Shaggy, peeling bark in large strips.
- Sweet edible nuts, tough shells.
- 5-part leaf arrangement.

Recipe featuring *Carya ovata*: Hickory Nut Butter (page 243).

CASTANEA DENTATA (American Chestnut)

FLOWERS AND FRUIT

- Flowers: Long catkins (male) and smaller female catkins; early summer.
- Fruit: Spiny burs containing 1 to 3 sweet chestnuts.

LEAF FEATURES

- Lanceolate, 5 to 9 inches, sharply serrated with forward-pointing "teeth."
- Shiny above, paler below.

STEM FEATURES

- Smooth, chestnut-brown twigs; small, pointed buds.
- Bark grayish-brown, shallowly fissured in maturity, but large trees are now rare due to blight.

GROWTH HABIT

- Historically a large canopy tree (60 to 100+ feet).
- May survive as a stump sprout; rarely reaches full maturity in the wild.

HABITAT AND GROWING CONDITIONS

- Well-drained, acidic forest soils.
- Native to eastern North America, once very common.

NOTABLE IDENTIFYING CHARACTERISTICS

- Spiny burs with sweet chestnuts.
- Long, serrated leaves with pointed teeth.
- Devastated by chestnut blight; large specimens are uncommon. However, restoration efforts are demonstrating increasingly promising results!

Recipes featuring *Castanea dentata*: None (since the American chestnut is functionally extinct); however, you can make Chestnut Flour and Chestnut Pancakes (page 245) using Chinese or Italian chestnuts.

CERIOPORUS SQUAMOSUS (Dryad's Saddle, Pheasant Back)

CAP

- Large, fan-/kidney-shaped, yellowish-tan with thin brown scales (ticked, pheasant feather pattern).

UNDERSIDE FEATURES

- Pore surface (no gills), large, angular pores at maturity.
- White to cream pores.

STIPE FEATURES

- Stubby, thick stipe; tough/fibrous in older specimens.

GROWTH HABIT

- Grows singly or in layered clusters on hardwoods.
- Spring to early summer.

HABITAT AND GROWING CONDITIONS

- On decaying logs/stumps (maple, elm).
- Cool, moist conditions, generally recurs annually in spring and fall.

SPORE PRINT

- White.

NOTABLE IDENTIFYING CHARACTERISTICS

- Cucumber/melon-like odor when fresh.
- Scaly cap resembling pheasant feathers.

Recipe featuring *Cerioporus squamosus*: Curried Pheasant Back Vegan White Chili (page 72).

CHENOPODIUM ALBUM (Lamb's-quarters, Goosefoot)

FLOWERS AND FRUIT

- Flowers: Greenish, in dense clusters at stem tips/axils.
- Fruit: Tiny seeds enclosed in papery covers, prolific seeder.

LEAF FEATURES

- Triangular to diamond-shaped, sometimes lobed edges.
- Whitish or powdery coating on young leaves, especially underside.

STEM FEATURES

- Erect, can be green to reddish, 1 to 3 feet tall (often taller).
- Branches with flower clusters.

GROWTH HABIT

- Fast-growing annual weed.

HABITAT AND GROWING CONDITIONS

- Disturbed soils: gardens, fields, roadsides.

NOTABLE IDENTIFYING CHARACTERISTICS

- Whitish, powdery undersides, inconspicuous green flowers.
- Common garden weed.

Recipe featuring *Chenopodium album*: Pulled Hen of the Woods with Juneberry Barbecue Sauce (page 266).

CICHORIUM INTYBUS (Chicory)

FLOWERS AND FRUIT

- Flowers: Bright blue petals with a fringed appearance.
- Fruit: Small achenes with a short bristly pappus.

LEAF FEATURES

- Basal leaves resemble dandelion (deeply lobed).
- Upper leaves smaller, more infrequent.

STEM FEATURES

- Stiff, wiry, can become woody at base; 1 to 4 feet tall.
- Often branched, with blooms along branches.

GROWTH HABIT

- Perennial herb, deep taproot.
- Roadsides, fields, disturbed soils.

HABITAT AND GROWING CONDITIONS

- Full sun, well-drained soil.
- Tolerates dry conditions.

NOTABLE IDENTIFYING CHARACTERISTICS

- Sky-blue "daisy-like" flowers situated on wiry, branching stems.
- Deeply lobed basal leaves.
- Roasted roots used historically as a coffee substitute.

Recipe featuring *Cichorium intybus*: Pulled Hen of the Woods with Juneberry Barbecue Sauce (page 266).

COLLYBIA NUDA (Wood Blewit)

(Also *Lepista nuda*, *Clitocybe nuda*)

CAP

- Lilac to purple when young, fading to tan with age, 2 to 5 inches wide.
- Convex to nearly flat, smooth.

UNDERSIDE FEATURES

- Margins may curl under (inrolled), then flatten out with age.
- Gills adnate or slightly decurrent, lilac turning pinkish with maturity.
- Crowded gills, frequently the most purple part of the mushroom.

STIPE FEATURES

- Sturdy, lilac color (fades to buff), may thicken toward base.
- Fibrous interior.

GROWTH HABIT

- Solitary or in groups on leaf litter, compost, or grassy areas.
- Late fall into early winter.

HABITAT AND GROWING CONDITIONS

- Saprotrophic in deciduous/mixed forests on decaying material.
- Prefers cool, moist conditions.

SPORE PRINT

- Pale, pinkish.

NOTABLE IDENTIFYING CHARACTERISTICS

- Purple/lilac coloration of cap/gills/stem (especially young).
- Slightly freezer-burned, fruity odor.
- Pale spore print.

Recipe featuring *Collybia nuda*: Half-Spicy Blewit Pasta (page 267).

COMPTONIA PEREGRINA (Sweetfern)

FLOWERS AND FRUIT

- Flowers: Inconspicuous catkins (male and female).
- Fruit: Small, burr-like clusters with seeds inside (nutlets), which may or may not be present.

LEAF FEATURES

- Narrow, deeply lobed leaves, fragrant when crushed.
- Fern-like appearance, 1 to 4 inches long.

STEM FEATURES

- Woody, branching stems; young twigs can be pubescent.
- Low shrub, 1 to 4 feet tall.

GROWTH HABIT

- Colony-forming via rhizomes, thrives on poor acidic soils.

HABITAT AND GROWING CONDITIONS

- Sandy, acidic sites, full sun to partial shade.
- Pine barrens, dry woods, disturbed areas. Frequently found near blueberries.

NOTABLE IDENTIFYING CHARACTERISTICS

- Strongly fragrant, deeply lobed leaves visually resembling fern fronds.
- Scrubby, gnarled branches, may form colonies.
- Dark green leaves in summer, turning coppery in autumn.

Recipe featuring *Comptonia peregrina*: Blueberry Goat Cheese Galette with Sweetfern (page 174).

CORNUS MAS (Cornelian Cherry Dogwood)

FLOWERS AND FRUIT

- Flowers: Bright yellow umbels in late winter/early spring.
- Fruit: Oblong red drupes, edible but tart, ripening mid-summer. Each berry contains a single oblong seed, pointed at each end.

LEAF FEATURES

- Opposite, oval leaves.
- 2 to 4 inches long, green turning subtle red/purple in fall.

STEM FEATURES

- Bark splits, revealing patches of brown/gray/orange on older trunks.
- Twigs have noticeable lenticels.

GROWTH HABIT

- Deciduous shrub/small tree, 8 to 15 feet (can reach 20+ feet).
- Dense branching, can be single or multi-trunked.

HABITAT AND GROWING CONDITIONS

- Ornamental in gardens/parks; well-drained soil, sun to partial shade.

NOTABLE IDENTIFYING CHARACTERISTICS

- Very early yellow flower clusters.
- Red "cherry-like" drupes.
- Colorful, mosaic-like bark on mature specimens.

Recipe featuring *Cornus mas*: Cornelian Cherry Fruit Leather (page 216).

CRATERELLUS SP. (Black Trumpet, Black Chanterelle)

CAP

- Trumpet-/funnel-shaped, thin-walled, deep hollow center.
- Dark gray to black exterior, sometimes wrinkled.

UNDERSIDE FEATURES

- Smooth or lightly veined, grayish black.
- No true gills, may have some wrinkling.

STIPE FEATURES

- Essentially continuous with the funnel; hollow interior.
- Tapers down.

GROWTH HABIT

- Scattered or clustered in leaf litter or moss, often partially hidden.
- Mid-late summer to fall fruiting.

HABITAT AND GROWING CONDITIONS

- Under hardwoods (especially oak/beech), moist/mossy areas.
- Likes well-drained forest floors, cemeteries.

SPORE PRINT

- Pale pink to orange, if procurable at all.

NOTABLE IDENTIFYING CHARACTERISTICS

- Dark funnel shape, hollow throughout.
- Subtle fruity/chocolate aroma.
- Thin-walled, may seem extremely brittle and dry or somewhat flexible/rubbery depending on conditions.

Recipe featuring *Craterellus* sp.: Lacto-Fermented Black Trumpets and Cultured Butter (page 146).

DAUCUS CAROTA (Wild Carrot, Queen Anne's Lace)

FLOWERS AND FRUIT

- Flowers: White, lacy umbels (2 to 4 inches across), sometimes with a tiny purple floret in center, though this is often absent. A lacy bract appears under the flower head, extending downward.
- Fruit: Spiny seeds forming a "bird's nest" shape as umbels contract in the fall.

LEAF FEATURES

- Finely dissected, fern-like leaves.
- Strongly herbal, carrot-like smell when crushed.

STEM FEATURES

- Erect, green, very bristly.
- Hollow or pithy interior.

GROWTH HABIT

- Biennial: first-year rosette, second-year flower stalk.
- Deep taproot (woody).

HABITAT AND GROWING CONDITIONS

- Fields, roadsides, disturbed areas, full sun.
- Well-drained soils.

NOTABLE IDENTIFYING CHARACTERISTICS

- Flat-topped umbels with lacy white flowers, folding into a "nest" at maturity.
- Skirt-like bract beneath flowerhead.
- Distinct carrot smell.

Recipe featuring *Daucus carota*: Thai Chanterelle Red Curry (page 148).

DESARMILLARIA CAESPITOSA (Ringless Honey Mushroom)

(Formerly *Armillaria tabescens*)

CAP

- Tan to brownish, 2 to 4 inches wide, sometimes scaly/fibrous.
- Convex to flat.

UNDERSIDE FEATURES

- Gills white to beige, attached to the stem.
- No ring on the stem.

STIPE FEATURES

- Slender, often fused at the base in clusters.
- Lacks the membranous ring typical of *Armillaria mellea*.

GROWTH HABIT

- Dense clusters on wood (stumps, roots, dying trees).
- Late summer to fall.

HABITAT AND GROWING CONDITIONS

- Saprotrophic/parasitic on hardwoods, warm climates.
- Often in large clusters.

SPORE PRINT

- White.

NOTABLE IDENTIFYING CHARACTERISTICS

- Clustered growth, no ring on stem.
- Honey-brown caps with subtle scales.

Recipe featuring *Desarmillaria caespitosa*: Honeyed Honeys (page 219).

DIOSPYROS VIRGINIANA (American Persimmon)

FLOWERS AND FRUIT

- Flowers: Small, greenish-yellow, bell-shaped (male/female on separate trees).
- Fruit: Round, golf ball–sized orange fruit with a star-shaped calyx, highly astringent unless fully ripe.

LEAF FEATURES

- Elliptical, 2 to 6 inches, glossy.
- Turn yellowish in fall.

STEM FEATURES

- Dark gray to black bark in rectangular blocks, like burnt charcoal.
- Twigs slender, brownish with small buds.

GROWTH HABIT

- Medium-sized tree.
- Fruit clings to branches well after leaves have fallen.

HABITAT AND GROWING CONDITIONS

- Old fields, forest edges, bottomlands, roadsides.
- Well-drained soils, sun to partial shade.

NOTABLE IDENTIFYING CHARACTERISTICS

- Blocky, thick bark; sweet orange fruit.
- Dioecious, typically need both male and female trees for fruit (excluding certain cultivars).
- Leaves with smooth margins, glossy above.

Recipe featuring *Diospyros virginiana*: Persimmon Bread (page 264).

ENTOLOMA ABORTIVUM
(Aborted Entoloma, Shrimp of the Woods)

CAP

- White, deformed, lumpy, puffy blobs.

UNDERSIDE FEATURES

- No distinct gills, irregular lumps, sometimes tinged with pink.

STIPE FEATURES

- Aborted lumps often lack a defined stem, but may come to a very short attachment point near or off the center.

GROWTH HABIT

- Late summer/fall in deciduous woods.
- Growing both on rotting wood and from the ground.

HABITAT AND GROWING CONDITIONS

- Often near or with *Armillaria* (honey mushrooms).
- Forest floor, leaf litter.

SPORE PRINT

- Pink.

NOTABLE IDENTIFYING CHARACTERISTICS

- Squashy, white lumps, frequently shaped like cocktail shrimp.
- Occasionally pinkish in the center.
- Growing low to the ground, often in groups, occasionally clustering somewhat, slight fishy smell.

Recipe featuring *Entoloma abortivum*: Spicy Shrimp of the Woods Onigiri (page 268).

EXSUDOPORUS FROSTII (Frost's Bolete)

(Previously *Boletus frostii* and then *Butyriboletus frostii*, but hopefully they don't change it again before this book comes out!)

CAP

- Bright red to crimson, sometimes sticky/slimy when fresh, 2 to 6 inches. Gelatinous top layer can be peeled away.
- Top layer of cap tastes citrusy when licked.

UNDERSIDE FEATURES

- Deep red pore surface, bruising blue but may appear black against red flesh.
- May ooze droplets of honey-colored guttation when fresh.

STIPE FEATURES

- Stout, red stipe with deep reticulation, may be accented with yellow along the edges.
- Flesh can bruise blue when cut.

GROWTH HABIT

- Solitary or scattered in hardwood forests (often oak).
- Summer to early fall.

HABITAT AND GROWING CONDITIONS

- Mycorrhizal with hardwoods, warm/humid regions. May be found near black cherry trees.
- Well-drained, rich forest soils.

SPORE PRINT

- Olive brown.

NOTABLE IDENTIFYING CHARACTERISTICS

- Vivid red cap, red pores, blue bruising, deeply reticulated stipe.
- Sticky cap tastes of lemon when licked.

Recipe featuring *Exsudoporus frostii*: Umami Bolete Dust (page 177).

FLAMMULINA VELUTIPES (Velvet Foot, Velvet Shank, Enoki)

CAP

- Small to medium, yellow-brown to orange-brown, often very sticky, resembling baked bread.

UNDERSIDE FEATURES

- Pale to white gills, sometimes darkening with age.

STIPE FEATURES

- Slender, velvety, dark brown.
- Upper portion is lighter, becoming darker toward the base.

GROWTH HABIT

- Grows in clusters on hardwood logs/stumps.
- Late fall to early spring fruiting.

HABITAT AND GROWING CONDITIONS

- Found in temperate regions on dead/dying hardwood.
- Prefers cool weather, often forms clusters under sloughing bark on dead tree trunks.

SPORE PRINT

- White.

NOTABLE IDENTIFYING CHARACTERISTICS

- Velvety stipes meeting at a common base.
- Bright, sticky caps reminiscent of baked bread in winter months.

Recipe featuring *Flammulina velutipes*: Miso Soup with Enoki (page 40).

GAULTHERIA PROCUMBENS (Wintergreen, Teaberry)

FLOWERS AND FRUIT

- Flowers: Small, white to pinkish, bell-shaped (mid-summer).
- Fruit: Bright red, pea-sized berries with a dimpled scar (persisting through winter).

LEAF FEATURES

- Elliptical, glossy, leathery, dark green, turning reddish in cold weather.
- Emits a strong wintergreen scent when crushed.

STEM FEATURES

- Slender, creeping stems with short, upright branches.
- Spreads via shallow rhizomes.

GROWTH HABIT

- Low-growing, mat-forming evergreen, creeping.

HABITAT AND GROWING CONDITIONS

- Common in coniferous/mixed forests, also often found with mature hardwoods.

NOTABLE IDENTIFYING CHARACTERISTICS

- Strong wintergreen aroma.
- Bright red berries persisting throughout winter.

Recipe featuring *Gaultheria procumbens*: Wintergreen Extract (page 39).

GRIFOLA FRONDOSA (Maitake, Hen of the Woods)

CAP

- Clusters of overlapping, oblong, fan-shaped fronds, forming a large rosette.
- May range in color from pale tan to dark grayish black.

UNDERSIDE FEATURES

- White pore surface.
- Pores small and tightly packed.

STIPE FEATURES

- Branching white stalk system from a thick base.

GROWTH HABIT

- Large clusters at base of hardwoods, primarily mature oak trees.
- Persists from early to late fall.

HABITAT AND GROWING CONDITIONS

- Saprotrophic/parasitic on living oak trees.
- Thrives during moist autumns in temperate forests.

SPORE PRINT

- White.

NOTABLE IDENTIFYING CHARACTERISTICS

- Multi-fronded, grayish-brown cluster resembling a hen's ruffled feathers.
- White pores beneath; thick branched base.
- Popular edible with earthy flavor.

Recipe featuring *Grifola frondosa*: Pulled Hen of the Woods with Juneberry Barbecue Sauce (page 266).

HOUSTONIA CAERULEA (Bluets, Quaker Ladies)

FLOWERS AND FRUIT

- Flowers: Tiny, pale blue to white, 4-lobed with a yellow center; spring/early summer.
- Fruit: Small capsule with seeds.

LEAF FEATURES

- Basal leaves small, with broad, rounded ends; stem leaves tiny, opposite.
- Forms low mats.

STEM FEATURES

- Slender, erect stems only a few inches tall.
- One flower per stem tip.

GROWTH HABIT

- Delicate perennial, 2 to 4 inches tall.
- Often forms patches.

HABITAT AND GROWING CONDITIONS

- Moist, acidic soils in meadows, lawns, open woods.
- Partial sun to light shade.

NOTABLE IDENTIFYING CHARACTERISTICS

- Very small pale-blue flowers with yellow "eye."
- Low, tufted habit.
- Not technically considered edible (nontoxic), but it's my favorite flower and I get to decide what goes in the book (perks of being the author).

HYPOMYCES LACTIFLUORUM (Lobster Mushroom)

(Parasitic fungus infecting *Russula/Lactarius*)

CAP

- Deformed host mushroom's cap; covered partially or entirely by a bright orange/red crust.

UNDERSIDE FEATURES

- Original gills are often distorted or obscured by orange calculus, but may appear as ridgelike structures.

STIPE FEATURES

- Also covered in orange "lobster shell" crust; shape may appear to resemble that of a lobster's tail.

GROWTH HABIT

- Solitary or in small clusters, summer/fall, often buried under thick duff.
- Parasitizes certain *Russula/Lactarius* mushrooms.

HABITAT AND GROWING CONDITIONS

- Mixed hardwood forests; moist soil. Often found under conifers.
- Can be found where host mushrooms occur.

SPORE PRINT

- White.

NOTABLE IDENTIFYING CHARACTERISTICS

- Bright orange exterior, firm, dense flesh.
- Unique parasitic deformation of host.
- Called "lobster mushroom" for color/appearance.
- Slightly fishy odor.

Recipes featuring *Hypomyces lactifluorum*: None—but try it out as a substitute for *Entoloma abortivum* in the recipe for Spicy Shrimp of the Woods Onigiri (page 268).

HYPSIZYGUS ULMARIUS (Elm Oyster)

CAP

- White to off-white or pale tan, convex/flattened.
- Smooth surface, edges may be inrolled initially.
- May crack as it ages.

UNDERSIDE FEATURES

- White gills, slightly decurrent or notched.
- Crowded gills.

STIPE FEATURES

- Central or slightly off-center stem, white/cream, solid.
- Can be bulbous at base.

GROWTH HABIT

- Solitary or in small clusters on hardwoods (often elm).
- Late summer to fall.
- Frequently grows out of living or dead tree cavities.

HABITAT AND GROWING CONDITIONS

- Saprobic on dead or living hardwood trunks.
- Often grows high in the trees.

SPORE PRINT

- White to buff.

NOTABLE IDENTIFYING CHARACTERISTICS

- White to pale tan cap, thick stem.
- Gills can be slightly decurrent.
- Sometimes confused with oyster mushrooms, but gills are not fully decurrent and specimens rarely cluster.

Recipes featuring *Hypsizygus ulmarius*: None, but can be used in place of *Cerioporus squamosus* in Curried Pheasant Back Vegan White Chili (page 72), as the texture can be very similar. Simply skip the instructions about pore removal.

JUGLANS NIGRA (Black Walnut)

FLOWERS AND FRUIT

- Flowers: Drooping male catkins; female flowers in small terminal clusters.
- Fruit: About the size of a pool ball, round green husk enclosing a hard, wrinkly nut. Kernel is oily and whitish, often encased in a papery brown covering. Strongly scented—turpentine and cleaning products.

LEAF FEATURES

- Pinnately compound, lanceolate leaflets with serrated margins.

STEM FEATURES

- Dark bark with deep furrows in a diamond pattern.

GROWTH HABIT

- Large deciduous tree, can be over 100 feet tall with a broad trunk.
- Wide crown.

HABITAT AND GROWING CONDITIONS

- Likes rich, well-drained soils in valleys/bottomlands.
- Full sun to partial shade.

NOTABLE IDENTIFYING CHARACTERISTICS

- Strong-smelling leaves/husks, black nut hull stains.
- Dark, deeply furrowed bark.
- Green fruits the size of a pool ball, strongly scented.

Recipes featuring *Juglans nigra*: Black Walnut Hummus (page 242). Can also be used in Lacto-Fermented Garlic Mustard Pesto (page 78).

JUNIPERUS COMMUNIS (Common Juniper)

FLOWERS AND FRUIT

- Flowers: No flowers.
- Fruit: Small, berry-like cones. May start out green or gray with a thick bloom, eventually ripening to bluish-black.

LEAF FEATURES

- Sharp, needle-like leaves in whorls of three; silvery-green or blue-green. Other species of *Juniperus* trees may develop scale leaves at maturity.

STEM FEATURES

- Young twigs green; older bark grayish-brown, shredded appearance.
- Often gnarled branching.

GROWTH HABIT

- Can be a low, spreading shrub or small tree.

HABITAT AND GROWING CONDITIONS

- Rocky/sandy, well-drained soils; full sun. Frequently found in landscaping.

NOTABLE IDENTIFYING CHARACTERISTICS

- Prickly needles in sets of three.
- Aromatic, berry-like cones.

Recipe featuring *Juniperus communis*: Juniper Ash Polenta Cake (page 42).

LACTARIUS INDIGO (Indigo Milkcap)

CAP

- Blue to bluish-gray, 2 to 5 inches, with concentric zones of color. Occasionally blotchy.
- Funnel-shaped with a depressed center.

UNDERSIDE FEATURES

- Gills are blue, bruising greenish.
- Exudes blue "milk" (latex) when cut.

STIPE FEATURES

- Similar color to cap, may also be blotchy.
- Can show greenish tints if bruised.

GROWTH HABIT

- Solitary or in small groups, on ground in forests.
- Summer/fall fruiting.

HABITAT AND GROWING CONDITIONS

- Mycorrhizal with hardwoods/conifers.
- Moist, shaded sites. Often growing under conifer duff.

SPORE PRINT

- Cream to pale yellow.

NOTABLE IDENTIFYING CHARACTERISTICS

- Unique blue color and bright blue latex.
- Gills bruise greenish.

Recipe featuring *Lactarius indigo*: Indigo Ice Cream (page 179).

LAETIPORUS CINCINNATUS (White-Pored Chicken of the Woods)

CAP

- Semicircular to fan-shaped shelves in rosettes, often found at tree bases, on fallen logs, or sprouting from buried wood.
- Salmon-pink to pale orange on top, with the lightest color typically toward the edges.

UNDERSIDE FEATURES

- Cream to white pores (distinguishes it from *L. sulphureus*).

STIPE FEATURES

- Lateral attachment to wood; thick, white flesh inside. No distinct stems.

GROWTH HABIT

- Clustered at or near tree bases (often oak).
- Summer to fall.

HABITAT AND GROWING CONDITIONS

- Saprotrophic/parasitic on hardwood roots/bases.
- Warm, humid conditions.

SPORE PRINT

- White to yellowish.

NOTABLE IDENTIFYING CHARACTERISTICS

- White pore surface, salmon-pink to creamy orange top.
- Rosette-like growth form.
- Best gathered when supple and chunky, before shelves begin to flatten out.

Recipe featuring *Laetiporus cincinnatus*: "The Rice" (page 214).

LAETIPORUS SULPHUREUS
(Sulphur Shelf, Chicken of the Woods)

CAP

- Bright orange/yellow shelves with wavy edges at maturity, often resembling candy corn.
- Overlapping "shelves" can form large clusters.

UNDERSIDE FEATURES

- Sulfur-yellow pores.

STIPE FEATURES

- Lateral attachment to wood, thick white to yellow flesh.

GROWTH HABIT

- Overlapping clusters on trunks/stumps/logs.
- Summer to fall.

HABITAT AND GROWING CONDITIONS

- Parasitic/saprotrophic on hardwoods (esp. oak).
- Likes warm, humid weather.

SPORE PRINT

- White to yellowish.

NOTABLE IDENTIFYING CHARACTERISTICS

- Bright orange top, yellow underside with tight, densely clustered pores.

NOTABLE IDENTIFYING CHARACTERISTICS (continued)

- Thick, fleshy shelves that flatten and fan out with age, often with wavy edges.

Recipe featuring *Laetiporus sulphureus*: "The Rice" (page 214).

LAPORTEA CANADENSIS (Wood Nettle)

FLOWERS AND FRUIT

- Flowers: Small, greenish-white clusters in leaf axils (mid-summer), monoecious.
- Fruit: Tiny lens-shaped seeds.

LEAF FEATURES

- Oval to lanceolate, serrated edges, stinging hairs on surface.
- Leaves arranged alternately (key distinction from stinging nettle).

STEM FEATURES

- Green, translucent, covered in tiny stinging hairs.
- 1 to 3 feet tall.

GROWTH HABIT

- Herbaceous perennial, forms dense patches.
- Spreads via short rhizomes.

HABITAT AND GROWING CONDITIONS

- Moist, rich wooded areas, often near streams.
- Prefers partial to full shade.

NOTABLE IDENTIFYING CHARACTERISTICS

- Stinging hairs on leaves/stems, alternate leaf arrangement.
- Found in shady, moist woodlands.

Recipes featuring *Laportea canadensis*: Nettle and Magnolia Tiramisu (page 108), Nettle Cake with Lilac Frosting (page 111).

LINDERA BENZOIN (Spicebush)

FLOWERS AND FRUIT

- Flowers: Clusters of small, pale yellow blooms in early spring.
- Fruit: Green, berrylike drupes maturing to bright red in late summer/ fall, oily, fragrant and yellowish inside when crushed.

LEAF FEATURES

- Elliptic to obovate, 2 to 5 inches, spicy/citrusy aroma when crushed. Yellow in autumn.

STEM FEATURES

- Greenish-brown twigs; scratching the bark releases a spicy fragrance.
- Bark is smooth and gray, dotted with pimply lenticels.

GROWTH HABIT

- Multi-stemmed shrub structure.
- Deciduous shrub, 6 to 12 feet tall, rounded form.
- Forms clumps, often under forest canopy.

HABITAT AND GROWING CONDITIONS

- Moist, rich woodlands, shady stream banks.
- Partial shade, acidic to neutral soils.

NOTABLE IDENTIFYING CHARACTERISTICS

- Aromatic leaves and twigs, showy red berries.
- Early spring yellow flowers.

Recipe featuring *Lindera benzoin*: Spicebush Shortbread (page 218).

MAGNOLIA × SOULANGEANA (Saucer Magnolia)

FLOWERS AND FRUIT

- Large, saucer-shaped tepals, bright pink; early spring before leaves appear. Warm, ginger scent.
- Cone-like fruit appears when blooms die off.

LEAF FEATURES

- Elliptical/obovate (narrower at base), up to 6 inches long, glossy green.

STEM FEATURES

- Twigs thick, with fuzzy, football-shaped flower buds.
- Smooth, grayish bark on older limbs.

GROWTH HABIT

- Deciduous small tree or large shrub, multi-stemmed, broad crown.

HABITAT AND GROWING CONDITIONS

- Ornamental in gardens/parks; prefers full sun to partial shade.

NOTABLE IDENTIFYING CHARACTERISTICS

- Showy, pink spring blooms, strong ginger scent.
- Fuzzy buds visible in late fall and early spring.

Recipes featuring *Magnolia x soulangeana*: Fermented Magnolia Bud Tea (page 107), Nettle and Magnolia Tiramisu (page 108), Fiddlehead Summer Rolls (page 74).

MATTEUCCIA STRUTHIOPTERIS (Ostrich Fern, Fiddlehead Fern)

FERTILE FRONDS

- Dark brown, rigid fronds in mid-summer, spore-bearing, persisting through winter.

STERILE FRONDS

- Tall, green, ostrich plume-like, 2 to 4+ feet.

LEAF FEATURES

- Long, feather-like sterile fronds with lanceolate outline.
- Thick vegetal fiddleheads in spring covered in flaky brown scales, each fiddlehead with a deep, U-shaped groove along its stem, similar to a celery stalk.

STEM FEATURES

- No true stem; fronds emerge in a circular "crown" from rhizomes.
- Central rhizome can form a clump or colony.

GROWTH HABIT

- Rhizomatous, forming dense stands in moist, shaded areas.
- Vase-shaped cluster of sterile fronds with fertile fronds in center.

HABITAT AND GROWING CONDITIONS

- Moist, rich soil in partial to full shade (stream banks, damp forests).
- Fiddleheads appear in spring.

NOTABLE IDENTIFYING CHARACTERISTICS

- Vase-like arrangement of tall, elegant sterile fronds.
- Dark, stiff fertile fronds.
- Thick, tightly curled fiddleheads in spring with deep, celery-like grooves.

Recipe featuring *Matteuccia struthiopteris*: Fiddlehead Summer Rolls (page 74).

MONARDA FISTULOSA (Wild Bergamot, Bee Balm)

FLOWERS AND FRUIT

- Flowers: Purple, tubular florets in dense, round heads.
- Fruit: Tiny, inconspicuous nutlets.

LEAF FEATURES

- Opposite, lanceolate leaves with serrated edges.
- Strongly minty/oregano aroma when crushed.

STEM FEATURES

- Square stems, light green to slightly purplish.

GROWTH HABIT

- Herbaceous perennial, branching near top.
- Spreads by rhizomes, forms loose clumps.

HABITAT AND GROWING CONDITIONS

- Meadows, prairies, roadsides, open woods.
- Full sun to partial shade, dry to moderately moist soil.

NOTABLE IDENTIFYING CHARACTERISTICS

- Showy purple flower heads with ragged, tubular petals.
- Aromatic, minty foliage, square stems.

Recipe featuring *Monarda fistulosa*: "The Rice" (page 214).

MORCHELLA AMERICANA (Yellow Morel, American Morel)

CAP

- Conical or egg-shaped with a honeycomb pattern of pits/ridges; yellow-tan to light brown. Other species may appear dark brown or gray/black.

UNDERSIDE FEATURES

- Same pitted surface continues around; the interior is hollow.

STIPE FEATURES

- Hollow, whitish to pale.
- Attached at the base of the cap (some related *Morchella* species may be detached).

GROWTH HABIT

- Solitary or in small groups, growing from the ground.

HABITAT AND GROWING CONDITIONS

- Near hardwoods (ash, elm, apple); moist, loamy soils.
- Springtime temperature shifts and moisture trigger fruiting.

SPORE PRINT

- Yellow to cream.

NOTABLE IDENTIFYING CHARACTERISTICS

- Entire mushroom is hollow.
- Highly prized edible with honeycomb-pitted cap.

Recipe featuring *Morchella americana*: Morel Risotto with Saffron, à la Debs (page 80).

MYOSOTIS ARVENSIS (Forget-Me-Not)

FLOWERS AND FRUIT

- Flowers: Tiny (2 to 5 mm) sky-blue with yellow/white eye.
- Fruit: Tiny nutlets in the calyx.

LEAF FEATURES

- Oblong to lanceolate, softly hairy.
- Basal leaves larger; upper leaves smaller.

STEM FEATURES

- Erect or ascending, covered in short hairs, typically branched near top.

GROWTH HABIT

- Annual or biennial herb.
- Forms low patches in fields/disturbed sites and by water.

HABITAT AND GROWING CONDITIONS

- Moist, disturbed soils, meadows, roadsides.
- Partial to full sun.

NOTABLE IDENTIFYING CHARACTERISTICS

- Tiny bright-blue flowers with a yellow eye.
- Fine hairs on leaves/stem.

Recipe featuring *Myosotis arvensis*: Wildflower Laminated Pappardelle (page 113).

PICEA ABIES (Norway Spruce)

FLOWERS AND FRUIT

- Mature cones (female) are large, cylindrical, hanging down.
- Male cones smaller, produce pollen in spring.

LEAF FEATURES

- Stiff, square, four-sided needles, dark green, each on a small woody peg.
- In early spring, new growth appears at the tips of each branch, bright vegetal green, lighter and much more tender than the older growth.

STEM FEATURES

- Pendulous branchlets (subdivided branches, especially toward terminal end) on larger branches.
- Bark grayish-brown, scaly.

GROWTH HABIT

- Tall, conical evergreen.
- Dense branching when young, drooping with age.

HABITAT AND GROWING CONDITIONS

- Planted ornamental and timber species; prefers cool climates, well-drained soil.
- Full sun, hardy and fast-growing.

NOTABLE IDENTIFYING CHARACTERISTICS

- Drooping secondary branches, long hanging cones.
- Rough branch texture from persistent leaf pegs.

Recipes featuring *Picea abies*: Mugolio (page 115), Spruce Tip Granita (page 117).

PINUS RESINOSA (Red Pine)

FLOWERS AND FRUIT

- Cones (female): Ovoid, 1½ to 2½ inches, brown when mature.
- Male cones small, clustered, shedding pollen.
- Seeds winged for wind dispersal.

LEAF FEATURES

- Needles in fascicles of two, 4 to 6 inches long, snapping cleanly when bent.

STEM FEATURES

- Bark develops reddish-brown plates with age.
- Branches in distinct whorls.

GROWTH HABIT

- Medium to large conifer, 50 to 80 feet (can reach 100+ feet).
- Straight trunk, rounded crown in older trees.

HABITAT AND GROWING CONDITIONS

- Prefers well-drained, sandy/loamy soils in full sun.
- Commonly planted for timber/reforestation in cooler climates.

NOTABLE IDENTIFYING CHARACTERISTICS

- Two-needle pine, each needle 4 to 6 inches, "snaps" cleanly.
- Reddish tinge to bark plates.

Recipe featuring *Pinus resinosa*: Mugolio (page 115).

PLEUROTUS CITRINOPILEATUS (Golden Oyster Mushroom)

CAP

- Bright yellow to golden and dulling with age, smaller and more brittle than *P. ostreatus*.
- Funnel-like, kidney-shaped caps, deeply dimpled.

UNDERSIDE FEATURES

- White to pale yellow gills, decurrent.
- Close, thin gills.

STIPE FEATURES

- White, off-center stems fused at a common base in clusters.
- Delicate texture, deeply curved.

GROWTH HABIT

- Invasive in North American forests. Often large, bouquet-like blooms on dead hardwood, especially elm and box elder.
- Summer/fall in warm climates; widely cultivated.

HABITAT AND GROWING CONDITIONS

- Saprotrophic; prefers warm, humid conditions.
- Invasive in mixed hardwood forests, often with multiple flushes per season.

SPORE PRINT

- White.

NOTABLE IDENTIFYING CHARACTERISTICS

- Golden-yellow caps clustered together in blooms.

NOTABLE IDENTIFYING CHARACTERISTICS (continued)

- Growing from dead or dying hardwood.
- Thin, delicate flesh compared to common oyster.

Recipes featuring *Pleurotus citrinopileatus*: Fried Oyster Mushroom Sandwich with Smashed Blackberries (page 211); can be used in place of enoki in Miso Soup with Enoki (page 40).

PLEUROTUS OSTREATUS (Oyster Mushroom)

CAP

- Oyster shell or fan-shaped, grayish, tan, or brownish. Fruit bodies may be substantial and are not typically flimsy or breakable.
- Smooth, inrolled edges when young.

UNDERSIDE FEATURES

- Cream/light gray gills decurrent down the stem.
- Gills densely packed.

STIPE FEATURES

- Often lateral or absent, occasionally multiple stipes meeting at a common point of attachment if present.
- White to grayish.

GROWTH HABIT

- Overlapping clusters on dead/dying hardwood.
- Year-round in mild climates, especially spring/fall.

HABITAT AND GROWING CONDITIONS

- Saprotrophic on deciduous wood.
- Moist, shaded forests with ample rotting wood. Unfussy, will thrive in many environments.

SPORE PRINT

- White to lilac gray.

NOTABLE IDENTIFYING CHARACTERISTICS

- Oyster-shaped cap, decurrent gills.

- White spore print.
- Mild odor/flavor.
- Often in layers on logs/stumps.

Recipes featuring *Pleurotus ostreatus*: Fried Oyster Mushroom Sandwich with Smashed Blackberries (page 211); can be used in place of enoki in Miso Soup with Enoki (page 40).

PODOPHYLLUM PELTATUM (Mayapple)

FLOWERS AND FRUIT

- Flower: White, waxy, ~2 inches, hidden under umbrella leaves (spring).
- Fruit: Lemon-shaped fruit ripening to yellow.

LEAF FEATURES

- Large, umbrella-like (peltate) leaves with 5 to 9 lobes.
- A single or double leaf per stalk; only double-leafed stalks flower.

STEM FEATURES

- Smooth, green stems 12 to 18 inches tall.
- Emerge from creeping rhizomes.

GROWTH HABIT

- Colonies of perennial herbs via rhizomes.
- Leaves die back by mid-summer, often before fruit has ripened.

HABITAT AND GROWING CONDITIONS

- Moist, rich deciduous woodlands, partial to full shade.
- Often in large patches.

NOTABLE IDENTIFYING CHARACTERISTICS

- Umbrella-like leaves, single hidden white flower.
- Edible ripe fruit, unripe fruit/leaves are toxic.
- Common in forest understories.

Recipe featuring *Podopyllum peltatum*: Mayapple Caramel Sauce (page 176).

PRUNUS AMERICANA (American Plum)

FLOWERS AND FRUIT

- Flowers: White, 5-petaled, in clusters on bare branches (early spring).
- Fruit: 1-inch round plums, a single stone in the center of each, becoming red or purple to yellowish when ripe in summer.

LEAF FEATURES

- Elliptic to oblong, finely serrated.
- Dark green above, paler below.

STEM FEATURES

- Reddish-brown young bark, older bark gray and scaly.
- Twigs may have small thorns/spurs.

GROWTH HABIT

- Small tree/large shrub, 10 to 20 feet tall, often thicket-forming.

HABITAT AND GROWING CONDITIONS

- Open woods, forest edges, stream banks, roadsides.

NOTABLE IDENTIFYING CHARACTERISTICS

- Early white blossoms, small sweet-tart plums.

Recipe featuring *Prunus americana*: Umeshu (page 141).

QUERCUS MACROCARPA (Burr Oak, Bur Oak)

FLOWERS AND FRUIT

- Flowers: Male catkins, small female spikes.
- Fruit: Large acorns with a fringed "burred" cup. The cup may cover the majority of the acorn.

LEAF FEATURES

- 6 to 12 inches long, rounded and deeply lobed with slender midsections.
- Dark green above, pale/fuzzy beneath.

STEM FEATURES

- Thick, ridged bark.
- Very stout branches.

GROWTH HABIT

- Large, broad-crowned deciduous tree, 70 to 80 feet or more.
- Fire-resistant bark.

HABITAT AND GROWING CONDITIONS

- Tolerates drought and periodic flooding.
- Found in prairies, savannas, bottomlands.

NOTABLE IDENTIFYING CHARACTERISTICS

- Huge, fringed acorns with oversized cups.
- Deeply lobed leaves with a waist.
- Thick, coarse bark and large trunk.

Recipes featuring *Quercus macrocarpa:* None.

ROSA RUGOSA (Rugosa Rose)

FLOWERS AND FRUIT

- Flowers: Large, fragrant, pink or white, 5 petals, bloom repeatedly in summer.
- Fruit: Plump, bright red/orange hips persisting into winter, typically the size of a large table grape.

LEAF FEATURES

- Pinnately compound with 5 to 9 leaflets, deeply wrinkled (rugose).
- Glossy green, serrated edges.

STEM FEATURES

- Dense, stiff thorns/bristles.
- Can form thick, impenetrable hedges.

GROWTH HABIT

- Deciduous shrub, 2 to 6 feet tall.

HABITAT AND GROWING CONDITIONS

- Coastal areas, dunes, roadsides, mall parking lots; very salt-tolerant.
- Full sun, well-drained soils.

NOTABLE IDENTIFYING CHARACTERISTICS

- Wrinkled leaves, numerous thorns.
- Fragrant flowers, prominent rose hips.
- Tough, hardy rose.

Recipes featuring *Rosa rugosa*: Rose Sugar (page 118), Spicebush Shortbread (page 218).

RUBUS ALLEGHENIENSIS (Allegheny Blackberry)

FLOWERS AND FRUIT

- Flowers: White, 5-petaled clusters (late spring).
- Fruit: Black aggregate drupelets (typical blackberry), mid-late summer.

LEAF FEATURES

- Compound leaves with 3 to 5 serrated leaflets.
- Undersides may be slightly hairy or thorny.

STEM FEATURES

- Arching canes with stiff thorns, may root at the tips.
- Green/reddish first-year canes; fruiting on second-year canes.

GROWTH HABIT

- Perennial shrub with biennial canes.
- Forms dense, thorny thickets.

HABITAT AND GROWING CONDITIONS

- Sunny edges, fields, roadsides.
- Fertile, moist soil but adaptable.

NOTABLE IDENTIFYING CHARACTERISTICS

- Robust thorns, white spring blossoms.
- Familiar blackberry fruit sprouting from canes.

Recipes featuring *Rubus allegheniensis*: Nettle Cake with Lilac Frosting (page 111), Fried Oyster Mushroom Sandwich with Smashed Blackberries (page 211).

RUBUS OCCIDENTALIS (Black Raspberry)

FLOWERS AND FRUIT

- Flowers: White, 5-petaled, spring.
- Fruit: Purplish-black raspberries with a hollow center, early summer.

LEAF FEATURES

- Compound leaves (3 to 5 leaflets), serrated, undersides often whitish, may have pubescent thorns.
- Smaller than blackberry leaves, often 3 leaflets.

STEM FEATURES

- Canes have a whitish coating of yeast bloom; smaller curved thorns.
- Arching canes may root at tips.

GROWTH HABIT

- Perennial shrub, biennial canes.
- Forms clumps or small thickets.

HABITAT AND GROWING CONDITIONS

- Forest edges, fields, roadsides, disturbed areas.
- Partial to full sun, well-drained soils.

NOTABLE IDENTIFYING CHARACTERISTICS

- Whitish bloom on canes, hollow fruit core.
- Thorns smaller and more flexible than blackberry.

Recipes featuring *Rubus occidentalis*: None (they never make it into a recipe at my house, because there's nothing I could make that beats eating them by the handful).

RUDBECKIA LACINIATA (Sochan, Green-Headed Coneflower, Cutleaf Coneflower)

FLOWERS AND FRUIT

- Flowers: Yellow rays drooping from green/yellow cone center (mid-late summer).
- Fruit: Cylindrical achenes within the cone.

LEAF FEATURES

- Deeply lobed leaves (3 to 7 lobes), "cutleaf."
- Lower leaves notably dissected.

STEM FEATURES

- Smooth, hollow stems, often reddish and U-shaped toward the base when young.
- Can reach 3 to 8 feet tall or higher.
- Young stems are flexible and smooth, and the ones growing from the center are particularly tender.

GROWTH HABIT

- Clumping perennial, spreads by rhizomes.
- Forms large patches in moist areas.

HABITAT AND GROWING CONDITIONS

- Moist meadows, stream banks, ditches, partial to full sun.
- Rich soils, ample water.

NOTABLE IDENTIFYING CHARACTERISTICS

- Drooping yellow petals, greenish central cone.
- Deeply lobed leaves, tall hollow stems.
- Native wildflower, popular in naturalized gardens.

Recipes featuring *Rudbeckia laciniata*: Wildcrafted Sofrito (page 71), Fiddlehead Summer Rolls (page 74), "The Rice" (page 214).

RUMEX CRISPUS (Curly Dock)

FLOWERS AND FRUIT

- Flowers: Greenish to reddish, in dense, branching clusters on upright stalk.
- Fruit: Triangular seeds enclosed by papery sepals.

LEAF FEATURES

- Lanceolate leaves with wavy/curly margins, up to 6 to 12 inches long.
- Basal leaves larger and curlier than upper leaves.

STEM FEATURES

- Erect, stiff stems 1 to 3 feet tall (or taller).
- Often turns reddish-brown with age.

GROWTH HABIT

- Perennial weed, deep taproot.
- Can form clumps.

HABITAT AND GROWING CONDITIONS

- Disturbed areas, fields, roadsides, wet meadows.

NOTABLE IDENTIFYING CHARACTERISTICS

- Wavy-edged leaves, tall seed stalks turning rusty brown.
- Deep taproot.
- Triangular seeds with papery covers.

Recipes featuring *Rumex crispus*: None (but try toasting the seeds in the oven and adding them to your granola or hot cereals).

SAMBUCUS NIGRA (Black Elderberry)

FLOWERS AND FRUIT

- Flowers: Large, flat-topped clusters of small, white, star-shaped blooms (late spring).
- Fruit: Dark purple-black berries in clusters, mid-late summer.

LEAF FEATURES

- Pinnately compound, lanceolate leaflets with serrated edges.
- Opposite arrangement on stems.

STEM FEATURES

- Young stems greenish and flexible, older bark light brown/gray with lenticels.
- White, pithy interior.

GROWTH HABIT

- Large shrub to a small tree.
- Suckers, forming clumps.

HABITAT AND GROWING CONDITIONS

- Moist, fertile soils (stream edges, wetlands).
- Sun to partial shade.

NOTABLE IDENTIFYING CHARACTERISTICS

- Umbels of white flowers, then black/purple berries.
- Opposite, compound leaves.
- Berries and flowers edible when properly prepared; raw berries slightly toxic.

Recipe featuring *Sambucus nigra*: Elderflower Honey (page 152).

SYRINGA VULGARIS (Common Lilac)

FLOWERS AND FRUIT

- Flowers: Large, fragrant panicles (lilac to purple, sometimes white), spring-blooming.
- Fruit: Elongated brown capsules splitting open to release seeds.

LEAF FEATURES

- Opposite, heart-shaped leaves, smooth margins.
- Medium to dark green on both sides.

STEM FEATURES

- Young branches greenish/reddish-brown; older bark gray.

GROWTH HABIT

- Deciduous shrub/small tree, 8 to 15 feet tall.
- Can sucker, forming dense clumps.
- Many stems often cluster from a single base.

HABITAT AND GROWING CONDITIONS

- Ornamental; prefers full sun.

NOTABLE IDENTIFYING CHARACTERISTICS

- Iconic fragrant purple flower clusters.
- Opposite, heart-shaped leaves.

Recipe featuring *Syringa vulgaris*: Nettle Cake with Lilac Frosting (page 111).

TARAXACUM SP. (Dandelion)

FLOWERS AND FRUIT

- Flowers: Bright yellow composite heads (ray florets).
- Fruit: Spherical seed heads ("puffballs") with individual seeds on fluffy parachutes.

LEAF FEATURES

- Basal rosette of deeply lobed leaves pointing back toward the center.
- Milky latex when cut.

STEM FEATURES

- Hollow, leafless stalk exudes milky sap.

GROWTH HABIT

- Herbaceous perennial with taproot.
- Common in lawns, spreads by wind-dispersed seeds.

HABITAT AND GROWING CONDITIONS

- Adaptable to lawns, fields, roadsides.
- Grows in sun and most soil types.

NOTABLE IDENTIFYING CHARACTERISTICS

- Bright yellow flower heads, turning into white fluffballs.
- Lobed leaves in a basal rosette; milky sap.

Recipes featuring *Taraxacum* sp.: Wildflower Laminated Pappardelle (page 113), Fiddlehead Summer Rolls (page 74), Fried Oyster Mushroom Sandwich with Smashed Blackberries (page 211), Pulled Hen of the Woods with Juneberry Barbecue Sauce (page 266).

TYPHA LATIFOLIA (Common Cattail)

FLOWERS AND FRUIT

- Flowers: Dense brown cylindrical female spike, with a slender, canary yellow male spike above (which disintegrates).
- Fruit: Fluffy seeds dispersed by wind once spikes break apart.

LEAF FEATURES

- Long, flat, lanceolate leaves 1 to 2 cm wide.
- Gray-green, often taller than the flower spike.

STEM FEATURES

- Upright, unbranched stems, 4 to 6 feet tall (sometimes taller).
- Pithy interior.

GROWTH HABIT

- Aquatic/semi-aquatic perennial forming dense stands via rhizomes.

HABITAT AND GROWING CONDITIONS

- Marshes, shorelines, ditches, shallow freshwater up to ~2 feet deep.
- Tolerates various nutrient levels.

NOTABLE IDENTIFYING CHARACTERISTICS

- Iconic brown "cattail" seed head.
- Broad leaves, tall stalks in wetlands.

Recipe featuring *Typha latifolia*: Cattail Pollen Truffles (page 142).

URTICA DIOICA (Stinging Nettle)

FLOWERS AND FRUIT

- Flowers: Small, greenish, in drooping clusters (male and female on separate plants).
- Fruit: Tiny, single-seeded fruits in hanging clusters.

LEAF FEATURES

- Opposite, ovate-lanceolate leaves with serrated margins, covered in stinging hairs.

STEM FEATURES

- Square-ish stems, also with stinging hairs.
- Can reach 2 to 6 feet tall.

GROWTH HABIT

- Perennial forming clonal colonies via rhizomes.
- Dense stands in nutrient-rich soils.

HABITAT AND GROWING CONDITIONS

- Moist, fertile areas, partial to full sun (edges of fields, riverbanks).
- High nitrogen soils.

NOTABLE IDENTIFYING CHARACTERISTICS

- Stinging hairs on leaves/stems.
- Opposite leaves with serrated margins.

Recipes featuring *Urtica dioica*: Nettle and Magnolia Tiramisu (page 108), Nettle Cake with Lilac Frosting (page 111).

VACCINIUM CORYMBOSUM (Highbush Blueberry)

FLOWERS AND FRUIT

- Flowers: White/pinkish, bell-shaped in clusters (spring).
- Fruit: Crowned berries, green at first then maturing to purple or blue with a powdery bloom, sweet-tart flavor.

LEAF FEATURES

- Elliptical leaves, 1 to 3 inches, smooth or finely serrated.
- Red to maroon fall color.

STEM FEATURES

- Multi-branched woody stems, grayish or greenish bark.
- New twigs may be green or red-tinged.

GROWTH HABIT

- Deciduous shrub, 6 to 12 feet tall, can form colonies.

HABITAT AND GROWING CONDITIONS

- Acidic, moist, well-drained soils (pH 4 to 5.5).
- Bogs, woodland edges, or cultivated sites.

NOTABLE IDENTIFYING CHARACTERISTICS

- Small blue-purple crowned berries.
- Woody shrubs with rough bark.
- Elliptical green leaves turning red in autumn.

Recipes featuring *Vaccinium corymbosum*: Wild Blueberry Ginger Cocktail Syrup and Gimlet (page 172), Blueberry Goat Cheese Galette with Sweetfern (page 174).

VIOLA ODORATA (Sweet Violet)

FLOWERS AND FRUIT

- Flowers: Deep purple (sometimes white), fragrant, 5 petals with a spur on the lower petal. Resembles a small butterfly.
- Fruit: Small capsule splitting into three parts.

LEAF FEATURES

- Heart-shaped with scalloped margins, forming a basal rosette.
- Leaves can be slightly fuzzy or smooth.

STEM FEATURES

- Leaves/flowers on separate stalks emerging from the rhizome.

GROWTH HABIT

- Low-growing perennial, forms groundcover.
- Spreads by runners and seeds.

HABITAT AND GROWING CONDITIONS

- Moist, rich soil in partial shade (woodlands, lawns, gardens).
- Can naturalize easily.

NOTABLE IDENTIFYING CHARACTERISTICS

- Distinct sweet fragrance.
- Purple, nodding, butterfly-shaped flowers with a spur.
- Heart-shaped leaves in basal arrangement.

Recipes containing *Viola odorata:* Violet Jelly Milk Tea with Violet Syrup (page 109), Wildflower Laminated Pappardelle (page 113)

About the Author

Gabrielle Cerberville is a celebrated foraging educator, community mycologist, and climate advocate whose high-energy, humor-laced videos have attracted almost two million followers. Known online as the "Chaotic Forager" and affectionately dubbed the "Internet's Mushroom Auntie," she leads keynotes, workshops, and guided forays across the United States, championing accessible, ethical relationships with wild food and fungi. Cerberville is also a PhD student at the University of Virginia Department of Music, where her research in the Composition and Computer Technologies program explores how sound can deepen our dialogue with the natural world. Her debut book invites readers to reimagine taste, place, and responsibility.